# Conquering Mathematics

## From Arithmetic to Calculus

# Conquering Mathematics
## From Arithmetic to Calculus

Lloyd Motz
and
Jefferson Hane Weaver

Plenum Press • New York and London

Library of Congress Cataloging-in-Publication Data

Motz, Lloyd, 1910-
    Conquering mathematics : from arithmetic to calculus / Lloyd Motz
  and Jefferson Hane Weaver.
        p.     cm.
    Includes bibliographical references and index.
    ISBN 0-306-43768-6
    1. Mathematics.    I. Weaver, Jefferson Hane.    II. Title.
  QA37.2.M68    1991
  510--dc20                                                         91-9329
                                                                       CIP

ISBN 0-306-43768-6

© 1991 Lloyd Motz and Jefferson Hane Weaver
Plenum Press is a division of Plenum Publishing Corporation
233 Spring Street, New York, N.Y. 10013

Printed in the United States of America

To Minne and Shelley

# Preface

We have designed and written this book, not as a text nor for the professional mathematician, but for the general reader who is naturally attracted to mathematics as a great intellectual challenge, and for the special reader whose work requires him to have a deeper understanding of mathematics than he acquired in school. Readers in the first group are drawn to mental recreational activities such as chess, bridge, and various types of puzzles, but they generally do not respond enthusiastically to mathematics because of their unhappy learning experiences with it during their school days. The readers in the second group turn to mathematics as a necessity, but with painful resignation and considerable apprehension regarding their abilities to master the branch of mathematics they need in their work. In either case, the fear of and revulsion to mathematics felt by these readers usually stem from their earlier frustrating encounters with it.

This book will show these readers that these fears, frustrations, and general antipathy are unwarranted, for, as stated, it is not a textbook full of long, boring proofs and hundreds of problems, rather it is an intellectual adventure, to be read with pleasure. It was written to be easily accessible and with concern for the mental tranquility of the reader who will experience considerable fulfillment when he/she sees the simplicity of basic mathematics. The emphasis throughout this book is on the clear explanation of mathematical concepts. Wherever proofs are needed for clarity, they are developed using the kind of reasoning one applies to the solution of problems in general. We have included such proofs not only to clarify the mathematics, but also to demonstrate that mathematical reasoning is not a special kind of reasoning that requires a special kind of mind but can be produced by any mature mind.

Since all mathematics is essentially arithmetic in various guises, we have emphasized the importance of a thorough understanding of arithmetic as the basis for understanding all other branches of mathematics. Accepting this as a truism, those who understand arithmetic, which we all do to some extent, can read this book with much more confidence, gratification, and a greater sense of accomplishment than they would otherwise. With this in mind, we have devoted the first two chapters of our book to the basic laws of arithmetic, since these laws apply to all branches of mathematics. The reader, mastering these laws, will be well prepared to understand and enjoy the chapters that follow.

Arithmetic can be fully understood only if the properties of numbers are known; therefore we discuss these in some detail, pointing out that numbers have two aspects: the ordinal (ordering) aspect, which leads to geometry, and the cardinal (quantitative) aspect, which leads to the traditional arithmetic. In the first two chapters we discuss the various categories [integers, fractions, irrational numbers (algebraic and transcendental), prime numbers, and complex numbers] into which our number system has been divided. These various subgroups of the number system are all governed by the same arithmetic laws, but they differ from each other in certain basic properties we describe and analyze; these differences make arithmetic extremely interesting. We conclude these two chapters with a discussion of the different bases (e.g., base 10) on which a number system can be constructed and with a discussion of logarithms.

In Chapter 3 we show that algebra is essentially the arithmetic of letters or symbols, pointing out that it enlarges and generalizes arithmetic, releasing it from the restraints imposed on it by the specific quantitative features that numbers introduce. Algebra thus carries us from a numerical logic to a pure logic of entities that may or may not represent real, concrete, physical objects or numbers. Indeed, algebras have been created that have their own rules of addition and multiplication. Thus, in the algebra of vectors, the law of addition is quite different from that of arithmetic, and in noncommutative algebras the product $ab$ of two factors $a$ and $b$ does not have to equal $ba$ (multiplication is noncommutative; the order of the factors in a product is important). Clearly, in a noncommutative algebra the factors of a product are not ordinary numbers but represent some kind of operation which must be performed in a definite order. In another algebra, called Boolean algebra, the elements may be abstract objects, such as logical propositions and sets of abstract entities or electrical networks. Boolean algebra is thus the algebra of pure logic. We do not discuss these esoteric algebras in our text but mention them here to indicate the vast wealth of algebra.

Although any algebra can be developed as a logical system of abstract concepts and propositions without reference to any kind of concrete representation, it can be understood more easily if it can be described by such a representation. We do this in Chapter 4 where we develop the basic algebraic concepts graphically, which leads one quite naturally to the function concept. The graph (the plot of the dependence of one quantity on another) shows the relationship between the two entities being plotted and thus defines a function. The function concept leads to the polynomial and to the algebraic equation, which we illustrate by the contour of an imaginary road, laid off in one direction only (east-west), but rising and falling with respect to sea level. This contour reveals all the properties of a functional relationship, no matter how complex it may be. It also gives us a simple understanding of the solutions of the algebraic equation associated with a polynomial if we picture the road as cutting sea-level points to the east or west of our position. The distance from this position (the origin of our graph) of the sea-level cutting point is a solution of algebraic equation; the number of such solutions that may exist depends on the complexity of the functional relationship. We discuss these from the simple to the complex.

The limitation of the number system to points on a line must be removed if we are to go from arithmetic to geometry. We do this in Chapter 5 where we introduce the concept of the dimensionality of a manifold. Limiting ourselves to the points on a plane—a two-dimensional manifold of points—we develop the geometry of triangles, which requires the introduction of the angle concept. We complete this chapter by showing how most theorems in geometry, such as the theorem of Pythagoras, can be deduced with elementary reasoning and a few geometric properties of triangles.

Continuing with geometry, we devote Chapter 6 to the geometry of the circle and to trigonometry. These two facets of mathematics go together because all the trigonometric properties of angles can be deduced from the unit circle, that is, a circle whose radius is one unit of length (one centimeter, one inch, one foot, etc.). This method of introducing trigonometry removes all mystery associated with it and permits the reader to grasp its basic elements quite easily. Since the circle is a very special case of a class of geometric figures—called conic sections—that play a very important role in physical phenomena from astronomy to atomic physics, as well as in mathematics itself, we describe the mathematics of conic sections in Chapter 7 which we call analytic geometry. This phase of mathematics, which was developed by Descartes, deals with the application of algebra to geometry; the most interesting results of this work are the algebraic equations of conic sections—the ellipse, the parabola, and the hyperbola. We derive these equations not by long algebraic and geometric arguments but from simple symmetry arguments, and then show, in a very elementary way, that the orbits of the planets are ellipses.

In the penultimate chapter, Chapter 8, we show how differential calculus arose simultaneously from Newton's study of the instantaneous speed of a body and Leibnitz's analysis of the rate of change of a function with respect to the entity on which it depends. We also discuss how the calculus made possible much of what we now know as modern science and technology.

Lloyd Motz
Jefferson Hane Weaver

# Contents

# CHAPTER 1

# The Number System

To see the great importance of numbers in our lives and daily activities, let us picture our affairs, if we can, if all numbers were erased from our minds and all numbers and all references to numbers were eliminated from wherever they are now found—such as streets, buildings, rooms, pages, bank accounts, and so forth. We would soon be in a constant state of confusion, uncertainty, and turmoil. Almost everything we do, and much of our thinking, whether we know it consciously or not, deals with or involves numbers. In a sense our lives are geared to numbers; we are often directed and controlled by numbers. Even our emotions respond (happiness or despair) to numbers. As stockholders, for example, we need only recall our elation or distress at substantial changes in the Dow–Jones average, even though most of us know little about the meaning of this number, to see how sensitive we are to

1

numbers. As taxpayers, we are quite aware of our anxiety when we are informed by the government that our tax records for the past several years are to be audited because we may believe we took a few deductions to which we were not completely entitled.

All of us are also acquainted with numbers that do not affect us directly or play important roles in our daily lives, but which we know to be important in nature's scheme of things or in history. Some examples of these pure numbers (stated without any description) are 186,000, 93 million, 200 billion, 24, 3.14159..., 365.25, 1776, 1914, 1939. These examples, most of which probably evoked images in the reader's mind, illustrate the many-faceted functions that numbers perform or the properties they have. Indeed, any one of these numbers can mean many things and we do not know precisely what the number really means until we attach an additional label to it even though it stirs our memory without such a label. The power and usefulness of numbers lie precisely in their flexibility, which permits us to apply them to cabbages, kings, pigs, and wings and, more or less, other useful things.

By attaching labels to the array of numbers above, we eliminate all uncertainty about them. Thus, "miles per second" attached to the first of these numbers identifies it as the speed of light, and "miles" attached to the second number marks it as the mean distance of the Earth from the sun. The number 3.14159 . . . is an approximation to a very important mathematical quantity called pi ($\pi$), first calculated by the famous Greek mathematician Archimedes. By specifying the last three numbers as dates, we relate them to very important events in world history. Inserting the word "stars" after the third number identifies our galaxy and "hours" after the fourth gives the day.

The concept of numbers has long fascinated mathematicians but the subject seems to have first received special attention from Pythagoras of Samos, whose followers held an almost mystical view of the power of mathematics. The Pythagoreans swore their members to secrecy; they devoted themselves to the study of philosophy, science, and mathematics at their school in the Greek colony of Croton in southern Italy. Because of the Pythagoreans' unorthodox social practices, however, they attracted the suspicions of their neighbors and were ultimately forced to leave the colony. Although their teachings were grounded in mysticism, "the Pythagoreans are credited with giving the subject of mathematics special and independent status. They were the first group to treat mathematical concepts as abstractions, and though Thales and his fellow Ionians had established some theorems deductively, the Pythagoreans employed this process exclusively and systematically. They distinguished mathematical theory from practices such as geodesy and calculation, and proved the fundamental theorems of plane and solid geometry and of *arithmetica*, the theory of numbers. To their dismay they also discovered and proved the irrationality of the square root of two."[1]

When many of us first learn about numbers, we are taught to use them to count classes of physical objects. Even though most persons can count, they would be hard put to define counting. There is nothing mystical about the nature of counting because it is nothing more than learning to compare classes of objects. In short, if we have two classes of objects, it is not necessary that we count all the objects in each class because we can simply arrange them in a one-to-one correspondence with each other and then see how many objects are left over. For example, if we know that a football stadium has 100,000 seats and we know that each seat is filled, then

we can conclude that there is a one-to-one correspondence
between the number of seats and the number of persons in
those seats without going through the laborious process of
counting each seat and each person in each seat. If there is
a standing-room-only section in the stadium to handle the
overflow crowds, we know that we can dispense with counting
the persons sitting in the 100,000 seats altogether and simply
concentrate on the number of persons standing in the section.
Since we could never be sure that all of these people would
be standing there at the time we did our counting, we would
probably check the ticket sales for the game and assume that
any tickets sold in excess of 100,000 would correspond to the
number of people in the "cheap seats."

Numbers are used on two levels, one of which (probably
the more important in our daily lives) is not associated with
an arithmetic and one of which is so associated. Here, by the
term "arithmetic," we mean a set of rules and axioms that
govern the operations with the numbers that are permitted.
On the first level are numbers such as Social Security numbers
or house numbers. These numbers play very important roles
in our lives—not from an arithmetic point of view but from
an informational one. You do not add, subtract, or multiply
these numbers because the result would be meaningless.
However, you use these numbers constantly to convey all
kinds of information to others about yourself without which
you would soon find yourself in nerve-wracking difficulties.
It is not difficult to imagine your consternation if you lose a
wallet or a purse with the items marked with these numbers
(e.g., credit cards, driver's license).

The arithmetic numbers in your life are your bank balance, the numbers you enter on your tax return, the numbers
that appear on the bills you receive and on the checks you
pay out. The arithmetic you apply to these numbers is gener-

ally no more than adding and subtracting with a bit of multiplication and division thrown in occasionally. When most people see a number, they respond to its quantitative character, which mathematicians call its cardinal aspect. If one sees the symbol 5, one immediately thinks of five things, but the ordinal aspect of a number, that is, its use to specify the position of an object in an array of similar objects, is at least as important as its cardinal aspect in mathematics. The ordinal aspect of a number may or may not be associated with its cardinal aspect. Thus, the number 9 on a house may or may not mean that it is the ninth house on the street, nor do we think of nine houses when we see the number 9 on a house; it is an identification symbol that enables us to locate the house on the street.

Although we shall consider both the cardinal and ordinal aspects of numbers in this book, we begin by examining the ordinal properties of numbers. To that end we lay off equally spaced points along a straight line, starting from any arbitrarily chosen point on the line. It does not matter which starting point on the line we choose since all points are identical. We may lay off the equally spaced points to the left of the chosen point or to its right; we begin by going to the right, and start out by labeling the chosen point with the symbol 0, which we may think of as zero or as the origin (the point from which we start).

The scale of our point-ordering scheme is set as soon as we choose the first point at some particular distance to the right of 0 and label it with the symbol 1. We may think of this symbol as identifying the first point in this ordering system to the right of 0 or as defining the unit distance measured from 0. Thus, the symbol 1 implies a distance-measuring operation which we now use to identify a point as far to the right of 1 as 1 is to the right of 0 and so on. Notice that where we place 1 to the right of 0 is immaterial to our ordering scheme, but once we choose this point, we can assign a symbol to every similar point to the right of 1 (similar in the sense that each such point is at the same distance from its right and left neighbors as 1 is from 0). With this idea in mind, we assign the eight symbols 2, 3, 4, 5, 6, 7, 8, 9 to the eight similar points to the right of 1. We thus exhaust the ten symbols that we grew up with from our childhood and which, we were taught, are the basis of our number system. Because this system contains exactly ten such symbols, we say that this number system has the base ten, and therefore call it a decimal system. All numbers are thus expressed as combinations of these ten symbols in the decimal number system.

Returning now to the ten points on our line that we have already labeled (notice that the point 9 is the tenth point, not the ninth point) we move on to the next point (the eleventh) to the right of 0 in this point-ordering scheme. How are we to label it and the points that follow it using only the ten symbols we have already introduced? Simply labeling the eleventh point 0, and the points that follow it 1, 2, 3, etc., would not do because this numbering system does not distinguish the second set of ten points from the first set, but we can make this distinction by placing the symbol 1 in front of each of the ten basic symbols. Thus, 10, 11, 12, 13, etc., label the eleventh, twelfth, thirteenth, etc., points in our numbering system and 19 labels the twentieth point. This is a very simple and elegant scheme which divides the points into groups of ten, with those in the first group labeled consecutively with the basic symbols 0, 1, 2, etc., and those in the second group labeled with the same symbols, in the same order, but with the symbol 1 in front of each one, this time around. We now follow this same procedure for the ten points following (to the right of) the twentieth point 19, but now we place the symbol 2 instead of the symbol 1 in front of the ten basic symbols of 0, 1, 2, etc. We thus label the third set of ten points 20, 21, 22, etc. By combining our basic symbols after 9 in sets of two (doublets, e.g., 12) we divide the first hundred points into ten sets of ten with 0 labeling the first point and 99 labeling the hundredth point in this scheme. Since we have exhausted all possible combinations in two of our basic symbols to label the first hundred points in our ordering scheme, we must use triplet combinations to label the second hundred points that follow the point 99. Thus, the hundred and first point is labeled 100, the hundred and second point 101, etc. Using sets of three basic symbols (triplets) such as 102, 110, 120, etc., we identify (label) the second hundred points, and see that triplet combinations (one hundred of them) beginning with the symbol 2, such as 200, 201, 202, encompass the third set of one hundred points. Finally, the tenth set of triplets, beginning with 900 and ending with 999, the thousandth point, labels the hundred points in the tenth set.

We can now go beyond (to the right of) the point 999 and use sets of four symbols such as 1000, 1001, 1200, etc., to label the second set of one thousand points and so on. If we continue this procedure, going to quintets, sextets, septets, octets, etc., of symbols, we obtain a nonending set of numbers which are called the positive integers. Zero (0) is the smallest positive integer but there is no single largest one.

The positive integers locate only the points (each of which is separated from its nearest neighbor on its left and right by the unit of distance) to the right of 0, but how do we take account, in this scheme, of the similar points to the left of 0? We use the same ten symbols but place a dash in front of each number, e.g., $-1$, $-2$, $-24$, $-1000$, etc., to tell us that the point it identifies is to the left of 0 rather than

to its right. Introducing the symbols with dashes in front of
them establishes a sense of direction on the line; if we start
from point 0 and want to move to a point whose position is
marked by a number with no dash in front of it, we move to
the right, but we move to the left if a dash is in front of the
number.

Mathematicians differentiate between the numbers
without a dash and those with a dash by calling the former
the positive integers and the latter the negative integers; the
integer 0 separates these two sets of integers, which, taken
together including 0, constitute the complete set of all integers.
Since these integers identify points on the line, and all such
points are identical, calling the integer that identifies a point
"negative" does not mean that this point has any kind of
"negative quality," but we shall see later when we speak of
the cardinal (quantitative) aspect of numbers that a negative
number will imply a quality different from that of a positive
number. To emphasize that there is no qualitative difference
between the integers as far as their ordinal aspect goes (in
using them merely to label different points or positions on a
line) we picture shifting all the integers on the line one step
to the right so that every integer is replaced by its left-hand
neighbor, e.g., $-2$ replaces $-1$, which replaces 0, which
replaces 1, which replaces 2, and so on. Nothing in this new
assignment of the integers to points on the line differentiates
it from the previous assignment.

We come now to the quantitative or cardinal aspect of
numbers which we all learned about in school and which
leads us to the idea of large numbers and small numbers. We
all think of any integer with many digits (the ten symbols
that are the base of our decimal number system) in it as larger
than a number with fewer digits in it. Thus, we think of the
integer 234 as being larger than the integer 12 (written sym-

bolically as $234 > 12$) even though there is not an iota of difference between the point identified by 234 and that point identified by 12. We arrive at this quantitative or cardinal difference between integers that we call large and those that we call small by introducing the first of the two arithmetic operations that govern arithmetic—addition.

We define the addition of two integers as the integer we obtain when we start from either one of these integers on the line and, moving to the right from that integer, count off a number of unit intervals (spaced points) equal to the second integer. Thus, the sum of 5 and 9 written as $5 + 9$ or $9 + 5$ is 14; that is, $5 + 9 = 9 + 5 = 14$. If we start from the point marked 5, we find that the ninth point following it is marked 14 and this is also the point we reach if we start from the point 9 and move five points to its right. The order in which we perform this addition operation does not affect the final result. We may interpret this rule of addition somewhat differently if we picture 5 as representing five unit steps from 0 and 9 as representing nine unit steps from 0. We may then say that starting from 0 we reach the point marked 14 either by taking five unit steps to the right followed by nine unit steps to the right or taking a single large step to the right equal to fourteen unit steps. In walking (or running) we often combine our steps in this way so our daily jaunts or walks carry their own adding machines with them which keep track of how many steps we take to cover various distances even though we may not be conscious of this addition. We do not have to be aware of the number of steps we take because we are guided by our sight, but a blind person keeps a very careful account of the sum of his footsteps to avoid becoming disoriented.

We may extend this concept of addition by applying it to more than two integers (e.g., the sum of five integers) such as $4 + 8 + 23 + 32 + 64$. Here, again, since the order in which we perform the addition does not alter the final result, we may rearrange these integers into groups of two or two and three if that simplifies the addition. We use parentheses to emphasize this grouping of the integers: $4 + 8 + 23 + 32 + 64 = (4 + 8) + (23 + 64) + 32 = (4 + 8 + 32) + (23 + 64) = (4 + 32) + (8 + 23) + 64$, etc. These examples of addition all involve positive integers or points to the right of 0, but to complete our definition we introduce negative integers (points to the left of 0) as well as positive ones. This operation leads us to the subtraction concept which is the inverse of addition. By picturing this operation in terms of points on the line, we see that a negative integer means moving from a starting point a given number of points to the left rather than to the right; the dash in front of an integer tells us to reverse our direction from right to left. Thus, the expression $5 + (-4) = 5 - 4$ means that we must start from point 5 and move four points to the left; the result, of course, is that we reach point 1 so that we have the ordinary definition of subtraction—we subtract four from five.

But we arrive at a somewhat deeper insight into the meaning of negative numbers by introducing the subtraction operation as displacement to the left along points on the line rather than as "taking away" one integer from another, as subtraction is usually defined. The difference between these two definitions of subtraction is most pronounced in a sum such as $5 + (-9) = 5 - 9$, which puzzles the reader who has not gone beyond what he learned in his elementary school days because the reader cannot understand how one can subtract a "larger number from a smaller one." But this sum presents no mystery if one interprets it as a sum of two different displacements along the line. Starting from 0 we move five steps to the right and then nine steps to the left so that we find ourselves four points to the left of 0 at −4. We could just as well have first moved nine steps to the left and then five steps to the right, again landing us at −4, so that the order in which we perform these operations is immaterial: $5 − 9 = −9 + 5 = −(9 − 5) = −4$.

The third expression in this series of qualities is particularly interesting because it demonstrates the usefulness of parentheses in presenting these sums. If we eliminate the parentheses, we must apply the minus sign (−) to each term inside the parentheses so that we obtain the expression $−9 − (−5) = −9 + 5$. This example illustrates the general rule that if a minus sign stands in front of a set of parentheses, we may bring the minus sign within the parentheses provided we change the sign of every term in the parentheses. Thus, $−(9 − 5 + 4 − 3) = (−9 + 5 − 4 + 3) = −9 + 5 − 4 + 3 = −(8 − 3) = −5$. This example illustrates another general rule that when two minus signs are next to each other, they cancel each other leaving a plus sign. If we have the negative integer −4 which tells us to move four steps to the left, the operation $−(−4)$ tells us to move four steps to the right so that $−(−4) = 4$.

Before we leave the addition of integers we introduce a simple rule for naming the very large ones which is based on representing such integers as sums of tens, hundreds, thousands, etc. The last digit in the integer is just the number of unit intervals in any set of ten points, the next to the last digit is the number of sets of ten unit intervals, the digit to the left of that is the number of sets of hundred unit intervals and so on, going to sets of a thousand, ten thousand, one hundred thousand, etc., as we move one digit to the left each time. Thus, the integer 9,763,548,493 is the number $3 + 90 + 400 + 8000 + 40,000 + 500,000 + 3,000,000 + 60,000,000 + 700,000,000 + 9,000,000,000$, expressed as 9 billion 763 million 548 thousand 4 hundred ninety three. A simple way to keep track of this nomenclature as applied to a large integer is to break it up into triplets of digits using commas, as shown in the above example.

Having discussed the addition of integers in detail we now turn to the multiplication of integers which we represent by an $\times$, a dot, or parentheses; the result of the multiplication operation is then called the product of the integers being multiplied. Thus, if 9 and 5 are multiplied we write $5 \times 9 = 9 \times 5 = 5(9) =$ the product of five and nine. From the point of view of using the integers that label the points on a line as described above we may define the multiplication of two integers as adding in equal groups the steps or unit intervals they represent. Thus, $5 \times 9 = 9 \times 5$ means that we add five groups of nine steps each or nine groups of five steps; we therefore also speak of the product as five times nine or nine times five. Starting from 0 we move nine steps to the right to point 9 and then nine more to the point 18 and so on, three more times, to reach the point 45. Or we can move from 0 five steps to the right to reach point 5 and so on eight more times to reach 45. The product of 5 and 9 or 9 and 5 is thus 45, which we obtain by either adding our steps from 0 in groups of nine or groups of five. Thus, multiplication, as defined here, is group addition.

Two special cases of multiplication, multiplying by 1 or by 0, warrant special treatment. The multiplication of any integer by 1 equals that integer so that $9 \times 1 = 1 \times 9 = 9$, which is obvious since this product means we are taking 1 nine times or nine once, which gives 9 in either case. The multiplication of any integer by 0 is always 0: $0 \times$ any integer $= 0 =$ any integer $\times 0$; no matter how many times we add 0 to 0 we get 0 and no matter how large an integer we take, we get zero if we take it zero times.

Multiplying by a negative integer also warrants special attention. We obtain the same numerical result as multiplying two positive integers but the product is itself negative so that the point it represents on the line is to the left of 0. Thus, $-9 \times 5 = -5 \times 9 = -(5 \times 9) = -45$; the negative sign changes the direction of our steps from right to left. If we multiply two negative integers, the product is the same as multiplying the same positive integers: $-5 \times -9 = (-5)(-9) = -(5)(-9) = -(-45) = 45$.

Multiplication is distributive which means that we can multiply a sum of any number of integers by multiplying each term in the sum by this integer and then adding all these products. Thus, as an example, $5(4 + 7 + 3) = 5 \times 4 + 5 \times 7 + 5 \times 3 = 20 + 35 + 15 = 5(14) = 70$. The product of an integer and a sum of integers equals the sum of the products of the integer and the separate integers in the sum. As another example involving negative integers, we have

$$-9(3 - 5 + 2 - 7) = -9 \times 3 - 9(-5) - 9(2) - 9(-7)$$

$$= -27 + 45 - 18 + 63$$

$$= 18 - 18 + 63 = 63 = -9(-7)$$

$$= 9(-3 + 5 - 2 + 7) = 9(7)$$

Note again that the minus sign in front of 9 changes the sign of each number inside the parentheses when $-9$ multiplies each of these numbers in turn.

We come now to division which is a form of subtraction, just as multiplication is a form of addition. We introduced multiplication by collecting the integers (or the points on the line they represent) into groups and then adding these groups. Since division is the inverse of multiplication, we introduce division by breaking up a large group of integers (or the points on the line marked by these integers) into a number of equal smaller groups and see how many such smaller groups the larger group contains. As specific examples, we consider the ten points 1, 2, 3, ..., 10 which define the ten unit intervals between 0 and 10. We can break these points into the five groups (1, 2), (3, 4), (5, 6), (7, 8), (9, 10) and say that ten divided by two is five. This operation is written as $10 \div 2 = 5$ or $10/2 = 5$ but the first expression, with the symbol $\div$ standing for division, is not used in mathematics, since it does not convey the sense of the division operation as well as the line does with one number on the top and one on the bottom. From here on we use the slash / to signify division. We can just as well divide 5 or any other number into 10. In dividing by 5 we have the two groups (1, 2, 3, 4, 5) and (6, 7, 8, 9, 10) so that $10/5 = 2$.

All of this is quite elementary and easy to follow if the number we divide by is such that we can break up the number we are dividing into a whole (integer) number of small groups. But what happens if we cannot do that operation? If we divided 10 by 3, we obtain the groups (1, 2, 3), (4, 5, 6), (7, 8, 9), 10. In other words, we obtain three groups of three, with one integer left over. We shall get to the meaning of that result when we introduce fractions but for the moment we say that 10 divided by 3 is 3, with one left over or $10/3 = 3$ with 1 as a remainder, which we write in our expression for division as $10/3 = 3 + 1/3$. We can easily understand this equation if we write 10 as $9 + 1$ and therefore write $10/3$ as $(9 + 1)/3$, or $9/3 + 1/3 = 3 + 1/3$. From this expression we see that division is distributive; if we have a sum of numbers and divide this sum by any number (the divisor) we may do it by first performing the sum and then dividing this number by the divisor, or we may first divide each number in the sum by the divisor and then sum these quotients (the quotient is the number obtained in the division of one number by another; thus, 3 is the quotient of 12 divided by 4). Thus, $(9 + 12 - 27)/3 = -6/3 = -2 = 9/3 + 12/3 - 27/3 = 3 +4 - 9 = -2$. We may also write this division example in various, but completely equivalent, forms:

$$(9 + 12 - 27)/3 = (1/3)(9 + 12 - 27)$$
$$= -(-9 - 12 + 27)/3$$
$$= (-1/3)(-9 - 12 + 27)$$
$$= (-9 - 12 + 27)/(-3)$$

The minus sign here must be understood as being equivalent to multiplication by $-1$. These examples tell us that if a sum of numbers is divided by a given number, we obtain the same quotient if we change the sign of the divisor and, at the same time, change the sign of every number in the sum. Just as the product of two negative numbers is positive, so too is the quotient of two negative numbers positive.

An important rule in division is that breaking up the divisor into a sum of numbers and then dividing the dividend (the number to be divided) by each of these numbers separately and adding these quotients does not give the desired quotient; such an operation is forbidden. Thus, the quotient of 48 divided by 12 or $48/12 = 48/(6 + 6) = 4$; it does not equal $48/6 + 48/6 = 8 + 8 = 16$. Division may not be performed by breaking up the divisor and dividing by these terms separately.

To show more clearly the relationship between multiplication and division we introduce the concept of the factor of a number. A factor of a given number is any integer, the product of which it and another integer is the given number.

Since $3 \times 4$ equals 12, the integers 3 and 4 are both factors of 12. Note that either factor is the dividend of the given number divided by the other factor or factors. This last statement is important because an integer may have more factors than one, as illustrated by the following expression: $12 = 3 \times 4 = 2 \times 6 = 3 \times 2 \times 2$. As a result, 12 has the factors 2, 3, 4, and 6, 6 has the factors 2 and 3, and 4 has the factors 2 and 2. A given number may have the same factor more than once; thus, 12 has the factor 2 twice, 8 has the factor 2 three times, 4 has the factor 2 twice, and 9 has the factor 3 twice ($9 = 3 \times 3$). Mathematicians have introduced a convenient shorthand notation to express this concept: The number of times a given factor is contained in an integer is shown by a superscript placed to the right and slightly above the given factor. Thus, $2^2$ means the factor 2 is contained twice in the integer, $2^3$ means it is contained 3 times, and so on, so that $12 = 4 \times 3 = 2 \times 2 \times 3 = 2^2 \times 3$;   $8 = 2 \times 2 \times 2 = 2^3$;   $9 = 3 \times 3 = 3^2$. Such superscripts, called exponents, direct us to multiply the integer by itself a given number of times (one fewer than the exponent). Thus, $2^2$ means multiplication of 2 by itself once, and $2^3$ means multiplication of 2 by itself, followed by the multiplication of that product (4) by 2. The exponent is also called "the power to which the integer is raised," so that $2^2$ means raising 2 to the second power (also called the square of 2) and $2^3$ means raising 2 to the third power (also called 2 cubed).

Exponents are a convenient shorthand notation for expressing very large numbers. One million may be expressed as 1,000,000 or $10^6$; one billion may be expressed as 1,000,000,000 or $10^9$, and so on. There is no limit to the size of the numbers that we can conceive of using our exponents. A favorite game of children when they learn about this powerful device is to raise a number to a very large exponent such as $10^{50}$ or $10^{100}$ or even $10^{500}$ which may be written out with a 1 followed by 50, 100, or 500 zeros, respectively. Surprisingly, the number $10^{100}$ is known as a googol, a name that was suggested by the mathematician Edward Kasner's nephew when his uncle asked him to name a number having one followed by one hundred zeros.[2] A googol is a very large number but it is not infinite. Although we will discuss infinite numbers in greater detail in a later chapter, we point out that many people confuse very large finite numbers with infinite numbers even though the two are fundamentally distinct. A finite number can be reached if we count long enough or add enough zeros but an infinite number can never be reached by such methods because it has been defined as being unbounded or without limit. We cannot simply write down the number one and then string billions upon billions of zeros behind it and expect to reach an infinite number if we persevere for a long enough time. Infinity is a quality that cannot be approached using our traditional number system. The German mathematician Georg Cantor began with the seeming paradox that an infinite class has the unique property that the whole is no greater than some of its parts and devised an elaborate hierarchy of infinite classes called transfinite numbers, the smallest of which describes the cardinality of the class of integers and is written with the Hebrew letter "aleph" (ℵ) followed by the subscript zero to refer to this smallest of transfinite numbers. Not surprisingly, an infinite number of transfinite numbers exists which are designated by the subscripts 0, 1, 2, etc. Because of the peculiar nature of transfinite numbers, however, they do not follow the same rules as finite numbers. For example, aleph-null plus any integer is equal to aleph-null; aleph-null multiplied by any integer is equal to aleph-null; and aleph-null raised to any exponent that is an integer is equal to aleph-null. We obtain the new transfinite number aleph-one (the letter aleph followed by the subscript 1) only when we raise aleph-null to the aleph-nullth power. This same operation is required of all other transfinite numbers in order to obtain a transfinite number with a higher cardinality. Even though the reader may be thoroughly bewildered by this idea of infinite classes and transfinite numbers, he is probably better off than our friend Cantor who was vilified or ignored by most of his contemporaries. He eventually went mad and more than a few people suggested that it was caused by his playing with his infinite classes too much.

Since we are discussing the usefulness of exponents for describing vast numbers in a convenient form, we consider some of the more prominent large numbers that have been used to represent physical quantities in the universe. We might think, at first blush, like the poets, that the stars in the sky or the grains of sand on the beach are infinite but they are actually finite in number. The number of grains of sand on a city beach probably does not exceed $10^{20}$ or one hundred quintillion or 100,000,000,000,000,000,000. The number of stars in the evening sky that may be observed with the naked eye is no more than a few thousand, but if our universe consists of at least one hundred billion ($10^{11}$) galaxies each having an average of one hundred billion stars, then the known universe has about $10^{22}$ or 10,000,000,000,000,000,000,000 or ten sextillion stars. On the atomic scale of nature, we encounter similarly large numbers when we talk about the number of atoms in an object. For example, an eyedropper holds about $10^{27}$ oxygen atoms, a number that is $10^5$ or 100,000 times larger than the number of stars we believe may exist in the known universe. Despite the incomprehensible vastness of these numbers, they pale into insignificance when compared with the googol or its progeny, the googolplex, which consists of a 1 followed by a googol of zeros. A googolplex may be written as $10^{googol}$ or 10 raised to the tenth power which is in turn raised to the hundredth power. "You will get some idea of the size of this very large but finite number from the fact that there would not be enough room to write it, if you went to the furthest star, touring all the nubulae and putting down zeros every inch of the way."[3] We can carry this mathematical game as far as we want and name our own very large but finite number (which we could christen the "supergoogolplex") by, for example, raising 10 to the googolplexth power. We might write this extremely large but finite number as $10^{googolplex}$ which would be more convenient than trying to write a googolplex of zeros after the number one. We could carry this nonsense even further by raising 10 to the googolplexth power which is in turn raised to the googolplexth power and so on without limit but this would become a fairly pointless exercise after a while because we have no known physical application for such huge numbers nor can they be easily comprehended. Moreover, it is not always a good idea to walk around muttering about raising a googolplex to the googolplexth power because such behavior will cause your family and friends to think you have lost your mind and encourage them to put you in a room with padded walls.

Large numbers obviously fascinate the public and they can often be used to gain the attention of persons who might not otherwise be inclined to lend an ear. The British astrophysicist Sir Arthur Eddington once began a lecture on cosmology by stating that there are exactly $136 \times 10^{80}$ protons in the universe and an equal number of electrons. It is doubtful that even the drowsiest members of Eddington's audience failed to sit up and take notice as he wrote out the number on the blackboard. Because there is no final arbiter, since we do not even know whether the universe has an edge or a finite number of stars, no one is really in a position to disprove Eddington's assertion. Based on our previous discussion, however, we know that Eddington's number is finite because it is smaller than a googol.

Although we have become more at ease in dealing with large numbers because of this marvelous mathematical concept of exponents, the Greek mathematician Archimedes may have been the first to suspect that some sort of mathematical shorthand could be used to represent very large numbers.

> "Archimedes contrived a procedure involving indices to reduce the problem to more wieldy dimensions. To the point he developed it Archimedes's system of *octads* yielded a number which in our notation would require 80,000 million million digits. This arithmetic could be elaborated (beyond the octads) to grind out numbers as large as desired, but the monster Archimedes had produced amply served him in his demonstration of the grains of sand it would take to fill the entire universe. Starting with the number of grains of sand that, placed side by side, would measure the width of a poppy seed, Archimedes presses forward with his exercise until he has the entire universe as stuffed as a child's beach pail."[4]

The sophistication with which Archimedes viewed the dichotomy between finite and infinite numbers was evidenced in the introductory passage of his essay *The Sand Reckoner*:

> "There are some, King Gelon, who think that the number of the sand is infinite in multitude; and I mean by the sand, not only that which exists about Syracuse and the rest of Sicily, but also that which is found in every region whether inhabited or uninhabited. Again there are some who, without regarding it as infinite, yet think that no number has been named which is great enough to exceed its multitude. And it is clear that they who hold this view, if they imagined a mass made up of sand in other respects as large as the mass of the earth, including in it all the seas and the hollows of the earth filled up to a height equal to that of the

highest mountains, would be many times further still from recognizing that any number could be expressed which exceeded the multitude of the sand so taken."

The number of factors in a large integer is, in general, larger than the number in a small integer, as illustrated by the integer 36, which has the factors 2, 3, 4, 6, 9, 12, and 18 in various product combinations such as $9 \times 4$, $6^2$, $3^2 \times 4$, $3 \times 12$, etc. Consider now the integer 32, which we may write as $2 \times 16 = 4 \times 8 = 2^2 \times 2^3 = 2^5$, which illustrates a very important arithmetic rule: to multiply two or more powers of the same integer, simply add all the exponents (the individual powers) and raise the integer to the power equal to that sum. Thus, $2^2 \times 2^3 \times 2^4 = 2^{(2+3+4)} = 2^9 = 512$. Note that the product of two different integers raised to the same or different powers cannot be obtained by adding the exponents (powers). The product $2^3 \times 3^5$ does not equal $(2 \times 3)^{(3+5)}$ nor $2^{(3+5)} \times 3$ nor $2 \times 3^{(3+5)}$. Adding the exponents is permitted only if the exponents apply to the same integer.

Since multiplying and dividing are the inverse of each other, the way we treat the two exponents involved when we divide an integer raised to a given power (exponent) by the same integer raised to the same or a different power is the inverse of what we do with these exponents when we multiply the two numbers. Thus, 32 divided by 8, that is, $32/8$, equals 4: $32/8 = 2^5/2^3 = 4 = 2^2 = 2^{(5-3)}$. Since this is equivalent to dividing $2^5$ by $2^3$ we see that in dividing one power of an integer by another power of the same integer, we obtain the quotient by subtracting the exponent of the divisor from that of the dividend. Whereas multiplication means adding exponents, division means subtracting them.

Returning now to $32/8 = 2^5/2^3$, we express this statement as a product by writing $2^5/2^3 = 2^{(5-3)} = 2^{[5+(-3)]} = 2^5 \times 2^{-3}$, noting that the difference $(5 - 3)$ may be written as the sum

$[5 + (-3)]$. We deduce from this statement that $2^{-3} = 1/2^3$, so that an integer raised to a negative power ($-3$ in this example) equals the reciprocal of the same integer raised to the same numerical positive power. From this fact we make a very interesting discovery about the value of any number raised to the power zero (0). We know that any number except zero divided by itself equals 1. For example, $5^3/5^3 = 1$. From what we have said above, this expression equals $5^3 \times 5^{-3} = 5^{3-3} = 5^0 = 1$. Since we may do the same thing with any number (other than zero), we see that any positive or negative number (except 0) to the zero power is 1.

We exclude 0 from this rule because dividing by zero is forbidden or, more precisely, is undefined (meaningless in arithmetic and, indeed, in all mathematics). Thus, such expressions as $0/0$, $1/0$, $0^0$ are undetermined in the sense that they may have any value or become infinite, depending on how we evaluate them. Of course, 0 multiplied by any number is zero.

We now extend the concept of the factor of a number by including itself and 1 among its factors, which leads us to the very important concept of a prime number, defined as any number that is divisible only by itself and 1 (it has only two factors: itself and 1). The integers 1, 2, 3, 5, 7, 11, 13, 17, 19, etc., are prime numbers. Finding all the prime factors of a number (expressing it as a product of prime numbers) is a very important exercise in arithmetic; or, as mathematicians express it, in the theory of numbers, which is one of the most difficult and demanding branches of mathematics. As examples of numbers expressed as products of their prime factors we have 12 ($3 \times 4 = 3 \times 2 \times 2$), 14 ($2 \times 7$), 70 ($2 \times 5 \times 7$), 210 ($2 \times 3 \times 5 \times 7$).

Do the prime numbers end? In other words, is there a largest prime number or do they go on forever? Some 2300 years ago Euclid (the discoverer of plane geometry) proved very easily that no largest prime number exists—that prime numbers go on forever. Euclid's proof essentially shows that the assumption that a largest prime exists leads to a contradiction. If we assume a largest prime number (call it P) exists and consider the number 1 plus the product of all the primes up to and including P [the number $(1 \times 2 \times 3 \times 5 \times \cdots \times P) + 1$], we see that this number is certainly not divisible by any of the primes from 1 up to P because dividing by any one of these primes leaves a remainder of 1. Therefore, either this number which is clearly larger than P is itself a prime, which means that P cannot be the largest prime, or this number is divisible by a prime not included in the product above and hence larger than P, which again means that P is not the largest prime number, contrary to our assumption.

Euclid is best remembered for his *Elements* in which he outlined his classical conception of a geometry of points and straight lines, relying totally on a fundamental set of postulates and axioms. Although we shall discuss Euclid's geometry in much greater detail in a later chapter, it is useful to point out here that his work failed to distinguish between the pure world of theoretical mathematics in which points are dimensionless and straight lines extend infinitely far in two directions and the real world in which a ruler allows us to draw a line that is not perfectly straight but will suffice for most purposes and a sharpened pencil permits us to make a point that is not really dimensionless even though it is very small. Euclid's work, like most subsequent works of theoretical or conceptual mathematics, was not concerned with the imperfections of the real world. His reasoning was based on rigorous logical proofs which in turn depended on his making certain assumptions about points, lines, planes, etc. In short, a work of pure mathematics such as Euclid's *Elements* has no choice but to rely on abstract mathematics or else its intellectual integrity would have been undermined. It was left to cartographers and engineers to deal with the discrepancies between Euclidean geometry and the less pristine physical objects observed in the real world whether they were the craggy coastline of Greece or the subtle imperfections of a brick's surface. "Evidently Euclid did not stress the practical aspects of his subject for there is a tale told of him that when one of his students asked of what use was the study of geometry, Euclid asked his slave to give the student three-pence, 'since he must needs make gain of what he learns.'"[5]

Eratosthenes, the great Alexandrian mathematician and geometer, who was the first to measure the diameter of the Earth in 250 B.C., described a simple systematic method for picking out the prime numbers from all the integers arranged in sequence 1, 2, 3, 4, 5, etc. Starting from 2 we cross out every second integer 4, 6, 8, 10, etc. Then starting from 3 we cross out every third integer 6, 9, 12, 15, etc., and then starting from 5 we cross out every fifth integer 10, 15, 20, 25, etc. We follow this procedure with each integer in turn that was not crossed out. We see that the numbers left are just the prime numbers 1, 2, 3, 5, 7, 11, 13, etc. because the numbers we cross out are those that can be divided by some smaller number and therefore are not prime numbers. Although this so-called "sieve of Eratosthenes" is an accurate way of pick-ing out every prime number, it is practical only for the small prime numbers. We can imagine the difficulty that Eratosthenes would have run into as he scratched long series of numbers in the sand and then ran back to cross out every second or third or fifth number before the wind obscured the markings or the tide came in. Since he could not count on the numbers in the sand lasting for more than a few days at best, he was naturally limited regarding how large a sequence of numbers he could analyze for their primes. More recently, mathematicians have devised various ingenious ways of determining whether a very large number is a prime number or not especially since prime numbers have begun to play an important role in cryptography.

A few more points about multiplication are worth noting. Consider the product $6 \times 8 = 48$. We may write this equation as $(4 + 2) \times (5 + 3)$ and multiply each number in the first parentheses by each number in the second and add them: $(4 + 2)(5 + 3) = 4 \times 5 + 4 \times 3 + 2 \times 5 + 2 \times 3 = 20 + 12 + 10 + 6 = 48$, so that the product of two sums of numbers is the sum of the products taken in every possible way. This statement is true for the products of two sums no matter how many terms each sum contains. Thus,

$$(2 + 1 + 3)(4 + 2 + 1) = 2 \times 4 + 2 \times 2 + 2 \times 1 + 1 \times 4 + 1 \times 2$$
$$+ 1 \times 1 + 3 \times 4 + 3 \times 2 + 3 \times 1$$
$$= 8 + 4 + 2 + 4 + 2 + 1 + 12 + 6 + 3$$
$$= 42$$

Note that every product in the final sum contains one factor from the first parentheses and one from the second combined in every possible way. Since the sum in each set of parentheses contains just three terms, the total number of terms in the final product is nine. An interesting application of this idea is to the sum and difference of the same two numbers such as

$$(9 - 8)(9 + 8) = 9 \times 9 + 9 \times 8 - 8 \times 9 - 8 \times 8 = 9^2 - 8^2$$
$$= 81 - 64 = 17$$

The sum $(9 + 24 + 16)$ is called a perfect square since it equals the square of a number, in this case 7 or the square of the sum of two numbers $(3 + 4)$. The general rule here is that the square of the sum of two numbers is the sum of the squares of the two numbers added to twice the product of the two numbers. This rule also applies to the two numbers if one is negative. Thus,

$$(4 - 5)(4 - 5) = (4 - 5)^2 = [4^2 + 5^2 + 2(4)(-5)]$$

$$= 16 + 25 - 40 = 1$$

Note that the square of a negative number is always positive; the general rule here is that if a negative number is raised to an even power, the result is positive, and if it is raised to an odd power, the result is negative. The product of two negative numbers is always positive (negative cancels negative) and the product of a negative and a positive number is always negative.

Since the number system described above is based on the ten distinct symbols (integers) $0, 1, 2, \ldots, 9$, we call it a decimal system and note that every number in this system can be expressed as a sum of powers of 10. Any single digit such as 4 can be written as $4 \times 1 = 4 \times 10^0$ (as a product of the digit and the zeroth power of 10). A three-digit number such as $534 = 5 \times 10^2 + 3 \times 10^1 + 4 \times 10^0$. We can write such a sum of integers multiplied by powers of 10 for any number, however large it may be. With this point understood we shall see later that we can represent any number we please as 10 raised to some power which is not, in general, an integer. We return to this important way of expressing any number later when we introduce the concept of the logarithm of a number.

We consider now the representation of numbers on a base other than ten such as a base which uses twelve symbols rather than ten. Suppose that the two additional symbols $\triangle$ and $\square$ were added to the symbols $0, 1, \ldots, 9$. The number ten would then be given by $\triangle$, the number eleven by $\square$, and the number twelve by 10. We would then again express our numbers as powers of 10, and $10^0$ would still be one but $10^1$ would be twelve, $10^2$ (100) would be one hundred forty-four, etc. The number eight would still be 8 or $8 \times 10^0$, but the number fifteen would be $3 \times 10^0 + 10^1$, so that on the base twelve, fifteen would be 13 and the number twenty-four would be $2 \times 10^1$ or 20 and twenty-five would be $2 \times 10^1 + 1$, or 21 and so on.

The base 2 is of particular importance in computers since only two symbols 0 and 1 are needed to express any number as powers of the base 2, which simplifies computations enormously. The numbers one, two, three, etc., on the base 2 are 1 (1), 10 (2), 11 (3), 100 (4), 101 (5), 110 (6), 111 (7), 1000 (8), 1001 (9), 1010 (10), where the base 10 expressions for these numbers are written in parentheses after their base 2 representations. To see how these numbers are expressed as powers of 2, we consider the last three numbers eight, nine, and ten. The number 8 equals $2^3 + 0 \times 2^2 + 0 \times 2^1 + 0 \times 2^0$. The coefficient of $2^3$ is 1 and the coefficients of the three other powers are zero. (The coefficient of a number raised to any power is the multiplier standing in front of it, which is written explicitly, unless it is one, in which case it is understood. The coefficients of the powers of 2 that express any number on the base 2 are either 0 or 1 and therefore these are the only digits that can be used to express any number on the base 2, which is why eight on this base is 1000.) The number nine equals $2^3 + 0 \times 2^2 + 0 \times 2^1 + 2^0$, hence its base 2 expression is 1001, and the number ten is $2^3 + 0 \times 2^2 + 2^1 + 0 \times 2^0$ so that its base 2 expression is 1010. Although this discussion of base 2 may seem somewhat academic, it is actually of great practical importance because all electronic computers carry out their calculations in base 2 notation because only two numbers (0 and 1) need to be used. Computers are like gigantic switchboxes and every single calculation involves tripping a number of switches from the "off" to the "on" position or vice versa. Because these switches exist only in the form of electronic circuits, we use the numbers 0 and 1 to tell the computer what procedures to follow with each calculation. In short, base 2 gives us the most manageable language for programming computers and for instructing them to carry out the most tedious and extensive calculations.

Despite the fear that many persons have that computers will one day run the world, we must remember that the computer is nothing more than a dumb counting machine. It must be fed a series of programming instructions before it can do any tasks. As a result, we should worry more about malevolent or dim-witted programmers taking over or at least mucking up the world instead of their electronic partners. "The great innovation of today's computers is their capacity to regulate themselves by digesting information which they themselves have produced. These machines cannot wonder, but they can respond. Having received elaborate and precise instructions as to the problem at hand, the computer proceeds to grinding out figures at a prodigious rate. In the course of this operation it in many ways apes the process of a human calculation. The machine can organize its problem into separate steps; it can use the results obtained in one step to execute the next; sometimes partial results are laid aside, that is, 'remembered,' while an intermediate step is carried through; trial and error methods are frequently used. Thus the computer guides itself by its own answers, makes choices, comparisons, decisions. It is less like a man than is an amoeba; nevertheless, it is more like a brain than any machine has ever been before."[6] However, we must remember that until someone actually switches it on and tells it to do something like transfer a company's assets to a secret Swiss bank account, the computer is essentially harmless. Short of one of today's larger mainframes toppling over onto a maintenance worker or perhaps electrocuting a programmer, it is hard to imagine how a computer could commit a "crime." The increasing numbers of programmers breaking into restricted access computer networks, however, suggests that some of their human programmers may have other ideas in mind.

We return now to the powers of 10 and note that we can use the exponents of 10 to represent the integers without writing down the entire power itself. We first write down only those integers that are the integer power of 10 and the powers of 10 after each such integer so that $1 = 10^0$, $10 = 10^1$, $100 = 10^2$, $1000 = 10^3$, $10,000 = 10^4$, etc. From these two types of numbers (those to the left of the equal sign and those to the right of the equal sign) which represent the same integers we see that we can label or represent each integer to the left of the equal sign by the corresponding exponent of 10 to the right of the equal sign. Thus, the integer powers 0, 1, 2, 3, 4 of 10 which increase by one (by addition) in successive steps, can be used to represent the integers 1, 10, 100, 1000, 10,000 which increase by a factor of 10 (multiplication) in successive steps. Note that no power of 10 can give us 0 so 0 cannot be represented in this way. In the 16th century the Scottish mathematician John Napier saw that the exponents of 10 (the powers of 10) can be used to simplify arithmetic by reducing multiplication to addition and division to subtraction as indicated above. He called the exponents of 10 that correspond to or represent the given integers the logarithms (logs) of these integers. Thus, 0, 1, 2, 3 are the logarithms of 1, 10, 100, 1000; since the logarithms are related to each other by simple addition and subtraction, whereas the numbers they represent are related by multiplication and division, working with logarithms lowers the level of the arithmetic involved. Thus, the multiplication of numbers can be replaced by the addition of their logarithms and the division of one number by another by the subtraction of their logarithms. Two simple examples illustrate this concept very nicely. If we consider the multiplication of 100 by 1000 ($100 \times 1000 = 100,000$), how de we represent this product using logarithms? The logarithm of 100 is 2, the logarithm of 1000 is 3, and the logarithm of 100,000 is 5, from which we see that $\log[(100)(1000)] = \log 100 + \log 1000 = 2 + 3 = 5 = \log(100,000)$.    On    the other hand, we see from the equation $100,000/100 = 1000$ that $\log(100,000/100) = \log\ 100,000 - \log\ 100 = 5 - 2 = 3 = \log 1000$, which means that in division the log of the dividend minus the log of the divisor equals the log of the quotient. We note in connection with this fact that the logarithm of any number (e.g., $100 = 10^3$) is the negative of the log of the reciprocal $1/1000$ of that number; that is, $\log 1000 = 3 = -\log 1/100 = -(-3)$.

Napier presented his concept of logarithms as a boon to mankind for, as he pointed out, working with logarithms reduces the drudgery of numerical computations enormously. Luckily, Napier came on the scene with his logarithms just when Johannes Kepler, the discoverer of the laws of planetary motion, was deeply immersed in mind-numbing, tedious calculations, filling hundreds of folio pages with lengthy arithmetic operations, in his construction of the orbit of Mars from the observational data of Tycho Brahe. To Kepler, this discovery was a gift from heaven, for logarithms reduce considerably the time he had to spend just doing arithmetic calculations, a task which he detested.

Some commentators such as Edward Kasner have suggested that Napier's publication of his book *A Description of the Admirable Table of Logarithms* in 1614 ranks second only to Isaac Newton's *Principia Mathematica* (which laid out the Newtonian conception of a mechanistic universe) as the high-water mark in British science.[7] Certainly, Napier's book required the intensity of concentration that had enabled Newton to conceptualize his remarkable vision of a universe governed by a pervasive law of gravitation; yet while Newton's book was densely worded and packed with tedious geometric axioms, it was probably much more enjoyable to compose (even for someone like Newton who disliked intensely the idea of publishing his ideas about motion and gravitation and thereby subjecting them to the critical scrutiny of his colleagues) than Napier's mind-numbing treatise which consisted of thousands of laborious calculations of the logarithms of numbers. Napier's work was certainly too detailed to have been an accidental discovery; he seems to have been motivated from the start by the desire to shorten the operations of multiplication and division for even the largest numbers. Florian Cajori recalls that Henry Briggs, who was then a professor of mathematics at Oxford University, "was so struck with admiration of Napier's book, that he left his studies in London to do homage to the Scottish philosopher. Briggs was delayed in his journey, and Napier complained to a common friend, 'Ah, John, Mr. Briggs will not come.' At that very moment knocks were heard at the gate, and Briggs was brought into the lord's chamber. Almost one quarter of an hour was spent, each beholding the other without speaking a word. At last Briggs began: 'My lord, I have undertaken this long journey purposely to see your person, and to know by what engine of wit or ingenuity you came first to think of this most excellent help in astronomy, viz., the logarithms; but, my lord, being by you found out, I wonder nobody found it out before, when now known it is so easy.'"

We now consider another important arithmetic operation which is the inverse of raising a number to an integer power such as $3^2 = 9$. In this operation we go from 3 to 9; the inverse of this procedure is going from 9 to 3, which we call taking the square root of 9, written as $\sqrt{9} = 3$. Other examples are $\sqrt{16} = 4$, $\sqrt{36} = 6$, but it is also clear that the square roots of most integers are not integers. Indeed, among all the integers from 1 to 100, only the integers 1, 4, 9, 16, 25, 36, 49, 64, 81, 100 have square roots that are integers. Later we shall consider how we are to represent noninteger square roots using the ten digits that are the basis of our decimal number system.

Note that the square root of any given number is that number which when multiplied by itself equals the given number. From this relation we obtain another way of writing the square root of a number. We have, taking $\sqrt{9}$ as an example, $\sqrt{9} \times \sqrt{9} = 9 = 9^1 = 9^{(1/2+1/2)} = 9^{1/2} \times 9^{1/2}$, since $1 = 1/2 + 1/2$, recalling that we add exponents when we multiply. Hence, $\sqrt{9} = 9^{1/2}$. Thus, the square root symbol $\sqrt{\phantom{x}}$ is equivalent to the exponent $1/2$. We shall return to this idea later in this chapter when we deal with fractions.

Just as we can have the square root of a number, we can have the cube root, the fourth root, etc., of any number written as $\sqrt[3]{\phantom{x}}$, $\sqrt[4]{\phantom{x}}$, etc. Thus, $\sqrt[3]{8} = 2$, $\sqrt[4]{16} = 2$, $\sqrt[3]{27} = 3$, $\sqrt[3]{64} = 4$. Here, too, the higher roots of most integers are not integers; only the integers 1, 8, 27, and 64 between 1 and 100 have integer cube roots and only the integers 1, 16 and 81 smaller than 100 have fourth roots that are integers. Note that all the roots of 1 are 1 since 1 raised to any power is 1.

Returning now to square roots, we have the following important rule (and this rule also applies to all the higher roots): The square root of a product of numbers is the product of the square roots. Thus, $8 = \sqrt{64} = \sqrt{(4 \times 16)} = \sqrt{4} \times \sqrt{16} = 2 \times 4$; also $\sqrt{144} = \sqrt{(16 \times 9)} = \sqrt{(4 \times 4 \times 9)} = 2 \times 2 \times 3$

= 12. This rule also applies to powers; the power of a product is the product of powers. Thus, $(3 \times 4 \times 2)^2 = (24)^2 = 3^2 \times 4^2 \times 2^2 = 9 \times 16 \times 4$. This rule does not apply to the square root of a sum of terms. Thus, $\sqrt{(9 + 16)} = \sqrt{25} = 5$ does not equal $\sqrt{9} + \sqrt{16} = 7$. The square of a sum does not equal the sum of the squares of the individual terms of the sum.

Working with powers and roots is greatly simplified by logarithms. We first consider, as an example, the power $4^3 = 4 \times 4 \times 4$; its logarithm, written log $4^3$, is the same as log $(4 \times 4 \times 4)$. But the logarithm of a product is the sum of the logarithms of the factors in the product. Thus, log $(4 \times 4 \times 4) = \log 4 + \log 4 + \log 4 = 3 \log 4$. We thus see that log $(4^3) = 3 \log 4$ so that the logarithm of the power of a number is the product of the power (in this example 3) and the log of the number itself. The same rule applies to roots. Thus, log $\sqrt{100}$ must equal log 10 or 1 since $\sqrt{100} = 10$. But log $\sqrt{100} = \log \sqrt{(10 \times 10)} = \log(\sqrt{10})(\sqrt{10}) = 2 \log \sqrt{10}$. But this expression can only equal 1 if log $\sqrt{10} = \log (10)^{1/2} = 1/2$. This is so if log $\sqrt{10} = 1/2 \log 10 = 1/2$. This expression shows us that the logarithm of the root of any number is the logarithm of the number divided by the root.

We see how the arithmetic of powers and roots of numbers, which are the inverse of each other, is equivalent to the multiplication and division (also the inverse of each other) of the logarithms by the powers of these numbers. Using logarithms, we replace multiplication by addition, division by subtraction, powers by multiplication, and roots by division. One more important aspect of logarithms is worth noting. Writing down the integers and their logarithms on the base 10 again, we have the integers followed in parentheses by their respective logarithms: 1 (0), 2 (?), 10 (1), 20 (?), 100 (2), 200 (?), 1000 (3), 2000 (?), 10,000 (4), etc. What about an integer, like 2, that lies between 1 and 10? Clearly its logarithm lies between 0 and 1 and can be expressed only by a number smaller than 1 which we discuss below. If we know the logarithm of 2 or any other such number, we can immediately write the logarithms of any numbers such as 20, 200, 2000, etc., which we obtain by multiplying 2 by 10, 100, 1000, respectively. We simply add 1, 2, 3, etc., to the logarithm of 2. Thus, $\log 20 = \log(2 \times 10) = \log 2 + \log 10 = 1 + \log 2$, or $\log 2000 = \log(2 \times 1000) = \log 2 + 3$.

## NONINTEGERS

We introduced the integers above by using the ten digits of our number system to label equally spaced points on a line. Since a point has no identity, we may start from any point we wish to present the ordinal properties of our number system, but we gain nothing by simply shifting all the integers any number of steps to the right or the left; everything remains just as it was originally. But we can enlarge our number system if we leave the integers on our line as they are and label points on the line that lie between the points labeled by two adjacent integers such as 0 and 1. Consider the point that lies midway between 0 and 1 and thus divides the interval into two equal parts. How shall we label it? We could, of course, introduce an entirely new symbol such as $\triangle$, but that would spoil our decimal system since we would have the eleven symbols, one of which ($\triangle$) would not be an integer. But we do not have to do this because we can use the ten digits we already have and combine them in an easily understandable way to label various points between 0 and 1. Returning to the midway point we note that it is the first in the set of points on the line, every other one of which divides one of the intervals following 0 into two equal parts; we therefore label this first point in a manner that expresses this idea for the first interval—namely 1/2. The number 1 which is called the numerator of this "fraction" simply identifies it as the first in a new sequence of numbers that can be used to label not only the points that are labeled by the integers 1, 2, 3, . . . but also all the points midway between any two integers. This new sequence of points is 1/2, 2/2, 3/2, 4/2, etc.

We note immediately that the point labeled 2/2 is the same as that labeled 1 by the second number in our integer sequence. We therefore write $1 = 2/2$; this expression shows us that a fraction equals the number we get when we divide the numerator of the fraction by its denominator (the number that is below the line of the fraction). This is to be expected since the word "fraction" itself means dividing a given quantity of anything into a sum of equal parts and taking one of these parts. In chemistry the word "fractionate" means separating a mixture of chemical compounds into its individual constituents or components but these are not necessarily equal in magnitude. If 100 grams of air, for example, were fractionated into its major components, 20 grams would be oxygen and 80 grams would be nitrogen (neglecting the small quantities of water vapor, carbon dioxide, ozone, etc.).

Returning now to the sequence 1/2, 2/2, 3/2, 4/2, . . . we see that 3/2 labels the point midway between 1 and 2; it may therefore be labeled, equivalently, as $1 + 1/2$ so that $3/2 = 1 + 1/2$, which we see immediately if we write 3/2 as $(2 + 1)/2 = 2/2 + 1/2$. This expression simply illustrates the basic rule that we may write any fraction as a sum of fractions by breaking up the original numerator into a sum of terms and placing each of these terms over the denominator of the original fraction and adding together all of these separate fractions, each of which, of course, is smaller than the original fraction. We see now that the fraction $4/2$ labels the point labeled by the integer 2 so that $4/2 = 2$. In the same way $5/2 = 2 + 1/2$, $6/2 = 3$, and so on. Using these fractions we label twice as many points on the line as we do when we use only the integers. Note that we may also label the midway points on the intervals to the left of 0 by introducing the negative fractions $-1/2$, $-2/2$, $-3/2$. . . .

We now pursue this idea further and divide the interval between 0 and 1 into three equal parts and use the fractions $1/3$, $2/3$, $3/3$, $4/3$, $5/3$, $6/3$, $7/3$, etc., to label all of these points in turn. We see at once that $3/3 = 2/2 = 1$, $6/3 = 4/2 = 2$, $4/3 = (3 + 1)/3 = 1 + 1/3$, $9/3 = 3$, and so on. From this expression we deduce the basic rule that if the numerator of a fraction equals its denominator, the fraction equals 1. This conclusion leads us to the very important rule that multiplying any fraction by a fraction whose numerator and denominator are equal does not change the value of the original fraction since this operation is equivalent to multiplying it by 1. Thus, as an example, $2/3 = 2/3 \times 1 = 2/3 \times 4/4 = (2 \times 4)/(3 \times 4) = 8/12$. Here we have used another very important rule: If we multiply one fraction by another we must multiply the two numerators together and the two denominators together to obtain the new fraction. We note that if two fractions have equal numerators but unequal denominators, the fraction with the *smaller* denominator is larger than the other fraction.

Proceeding as above we can divide the interval between points 0 and 1 into any number of equal parts we wish (4, 5, 6, etc.) and obtain the points labeled $1/4$, $1/5$, $1/6$, . . . $1/10$, . . . which lie closer and closer to 0. Each of these points establishes a new sequence of fractions such as $1/4$, $2/4$, $3/4$, $4/4$, . . . which label new points in each interval between any two integers and relabel those already labeled by previous sequences. Thus, 2/4 and 1/2 label the same point as do the fractions 3/6, 4/8, 5/10, 6/12, etc., and 1/3, 2/6, 3/9, 4/12 all label the same point (different, of course, from the point 1/2). Since our numbers are based on a decimal system (ten different digits) we pay particular attention to fractions that have 10 or powers of 10 as their denominators and we introduce a special shorthand way of writing such fractions, which eliminates the need to use the denominator 10. Starting with 1/10 we write it as .1 or 0.1 so that the dot placed in front of 1 stands for the denominator 10. This same shorthand notation is used for any fraction with the denominator 10 if the numerator is any integer between 0 and 10. Thus, $7/10 = 7 \times 1/10 = 7 \times 0.1 = 0.7$ or .7. Consider now the fraction 13/10. This fraction equals $13 \times 1/10$ or $13 \times 0.1$ equals $(10 + 3)/10 = 1 + 3/10 = 1 + 0.3 = 1.3$. This expression is an example of the decimal representation of fractions with 10 as their denominators. The general rule for expressing fractions with denominators of 10 as decimals is to place a dot in front of the last digit in the numerator. Thus, $3/10 = 0.3$, $24/10 = 2.4$, $89/10 = 8.9$, $125/10 = 12.5$, etc. Since a fraction is merely another way of representing the division of one number by another (the numerator by the denominator), we can represent the division of 10 into any integer as a decimal by simply placing a dot in front of the last digit in the integer. The quotients obtained by dividing 10 into the integers 5, 25, 125, 1251, etc., are 0.5, 2.5, 12.5, 125.1, etc.

Consider now the fraction 1/100 which is the product of the two fractions $1/10 \times 1/10$, also written as $1/10^2$ or $10^{-2}$, when we use the negative exponent to represent the reciprocal of the fraction 1/100. We may also describe the fraction 1/100 as the quotient of the fraction 1/10 divided by 10. But 1/10 is 0.1 and if we divide this quantity by 10 we must write it as 0.01 or .01 so that 1/100 is 0.01 written as a decimal, which leads to the following general rule: Any fraction with 100 as its denominator and an integer as its numerator can be written as a decimal by placing a dot in front of the last two digits of the integer in the numerator. This may at first appear puzzling, if not meaningless, when the integer in the numerator consists of a single digit such as 7 giving the fraction 7/100. What does placing a dot in front of the last two digits mean if we have only one digit? We give meaning to this idea by noting that the integer 7 can be written as 07, 007, 0007, etc., since a zero in front of an integer does not change its value. Placing a dot in front of the last two digits then gives us the decimal equivalent 0.07, or 7/100 (seven hundredths). In the same way we obtain the decimal equivalent of thousandths by placing a dot in front of the last three digits, and so on. For example, 12.6 means twelve and six tenths $(12 + 6/10$, or $12 + 3/5)$, 1.26 means one and twenty-six hundredths $(1 + 26/100$ or $1 + 13/50)$, and 0.126 means one hundred and twenty-six thousandths $(126/1000$ or $63/500)$.

The decimal representation of fractions is used extensively in the world of numbers as they relate to our daily activities. Our coinage (currency) and, hence, our pricing system are decimal systems, with the name of the unit coin, the cent, derived from the Latin word, centum, meaning hundred. One cent, of course, is one hundredth $(1/100)$ of a dollar and twenty-five cents is 25/100 or 1/4 of a dollar and, therefore, properly called one quarter and written $0.25 = 25¢$. Another application which we constantly deal with is the concept of percent, which means a fraction of one hundred. Thus, one percent means $1/100 = 0.01$ and 20 percent or 0.20 means $20/100 = 2/10 = 1/5$. This concept is generally used to express a given fraction of a quantity of some entity or other such as the percent of a given component or constituent in a mixture of various components; thus, oxygen is 20 percent of the atmosphere and nitrogen is 80 percent written as 20% and 80%, respectively. These quantities are equivalent, of course, to 1/5 (oxygen) and 4/5 (nitrogen).

Because a percent is a fraction of 100, it is often spoken of as a rate as in the "unemployment rate" which is the percentage (or fraction) of the entire work force that is unemployed. The widest use of the percentage concept is in expressing interest on bank deposits, on loans, on bonds, on income (tax), on mortgages, etc. Since interest on bank deposits is important to all of us, a simple example illustrates the essential feature and how it is calculated. Consider a $100 deposit (called the principal) in a bank which pays an interest of 6 percent per year. The "per year" is very important here because interest is understood to be a rate, which means the accumulation of money in a given period of time (in this example, one year) at a specified percentage. The interest rate can be specified as a percentage per year, per 6 months, per 3 months, per month, per day, or per second. As the time interval is reduced, so too is the percentage, and by exactly the same fraction as the time interval itself. Thus, 6% per year means 3% per 6 months (per half year or semiannually), 3/2% or 1.5% per 3 months (quarterly), 1/2% or 0.5% per month, and 6/365% per day. The interest earned in the first period is added on to the principal (the initial deposit) at the end of the first period, and the interest in the second period is calculated based on the principal plus this interest, and so on for each succeeding interval. This procedure is called compounding interest (interest on interest).

Returning now to the simple example of $100 earning interest at 6% per year, we see that the increase in the deposit after one year is $6/100 \times \$100 = 0.06 \times 100$ or $6 so that the new principal is $106. After the second year the principal is increased by $106 \times 6/100 = 1.06 \times 6 = 6.36$, becoming $112.36. If we compound the interest quarterly (4 times a year) we obtain for the principal after the first quarter $100 + (1.5/100) \times 100$ or $100(1 + 0.015)$. At the end of the second quarter the principal becomes

$$100(1 + 0.015) + 1.5/100 \times 100(1 + 0.015)$$
$$= 100[(1 + 0.015) + 0.015(1 + 0.015)]$$
$$= 100(1 + 0.015)(1 + 0.015)$$
$$= 100(1 + 0.015)^2$$

All we have done here is to use our multiplication rule which tells us that a given number (multiplier) multiplies each term in a sum. We may sum the terms first and then apply the multiplier, which we did with 100. We then did the same thing with the multiplier $(1 + 0.015)$. We thus obtain the following general rule, regardless of the size of the principal or the interest rate, if we wish to calculate the final principal (the original deposit plus the accrued interest) after compounding the interest a given number of times: Add the interest to 1 and raise this number to a power equal to the number of times the interest is compounded, then multiply this amount by the initial principal. Applying this procedure to the example discussed above of 6% interest per year, we find that after ten years the final principal is $100(1.06)^{10}$ which is 100 multiplied by 1.06 ten times and so on as we increase the number of years.

The procedure leads us quite naturally to another aspect of compounding interest. Suppose that we wish to calculate our principal at the end of a year if the interest is compounded a very large number of times per year such as a million times or ten million times or even an infinite number of times. How do we proceed? To make things very easy, let us suppose that the yearly interest is 100%, that we compound it 100 times, and we start with a principal of $1.00. The interest rate per hundredth of a year (100 equal intervals) is 1/100 and, following our rule, we find that at the end of the year our dollar has become $1(1 + 1/100)^{100} = (1 + 1/100)^{100} = (100/100 + 1/100)^{100} = (101/100)^{100}$. Of course, if the interest were not compounded, our dollar would become $2 at the end of the year, which is smaller than what we obtain by compounding 100 times, but not much smaller. "It seems clear that the more often the interest is compounded, the more money you will have in the bank. By a further stretch of the imagination, you may conceive of the possibility that the philanthropic bank decides to compound the interest continuously, that is to say at every instant throughout the year. How much money will you then have at the end of the year? No doubt a fortune. At least, that is what you would suspect, even allowing for what you know about banks. Indeed you might become, not a millionaire, not a billionaire, but more nearly what could be described as an 'infinitaire.' Alas, banish all delusions of grandeur, for the process of compounding interest continuously, at every instant, generates an infinite series which converges to the limit $e$."[8] The actual value of $e$ is discussed below.

If the interest is compounded continuously (for an infinite number of times) the interest in each of these infinitesimal intervals is infinitesimal, but the number of factors that multiply the initial principal (in this example $1.00) is infinite; the result, however, is a definite number, which can only be written as an infinite sum of rapidly decreasing (in size) fractions, namely $1 + 1 + 1/2 + 1/(2 \times 3) + 1/(2 \times 3 \times 4) + 1/(2 \times 3 \times 4 \times 5) \cdots$. Although the number of terms in this sum is infinite, the sum is not infinite. However, its finite value cannot be written down exactly because it is not an integer or a fraction. Mathematicians therefore represent it by its own symbol, the letter $e$, whose approximate value is 2.718281828. Since $e$ is neither an integer nor a fraction, it is placed in a special category of numbers called transcendental numbers to which the number $\pi$ (pi) also belongs. We discuss such numbers (which are part of a larger group called irrational numbers) further on noting here only that $e$ and $\pi$ play an enormous role in the structure of matter and the laws of nature as well as in statistics.

Continuing with the discussion of the properties of fractions we consider the addition of fractions, their subtraction, their multiplication, and their division which we discussed briefly above. To understand these arithmetic operations we note first that a fraction with any given denominator can be changed into an equivalent fraction with any different denominator we may desire. As an example we consider the fraction 1/3 and change it to a fraction with 5 in the denominator. We multiply 1/3 by 5/5 which leaves the value of 1/3 unchanged because 5/5 is 1. Thus, $1/3 = 1/3 \times 5/5 = 5/3 \times 1/5 = (5/3)/5$. Thus, 1/3 equals 5/3 fifths. Here the numerator of our fraction is, itself, a fraction. We may, of course, write this quantity as $5/(3 \times 5) = 5/15$. These are all equivalent ways of writing the fraction 1/3. This leads us to the addition of fractions which involves changing the denominators of the fractions to be added if they are different. If the denominators of these fractions are the same, we simply add the numerators and place this sum over the common denominator. Thus, $1/5 + 2/5 + 4/5 = 7/5$. Here we are adding quantities of fifths together which gives us a total number of fifths as the sum. If we are dealing with a sum like $1/3 + 2/5$, we may not add the numerators because the denominators are different, but we may do so if we first rewrite both fractions as fractions with the same denominator (we use a common denominator for both of them). We may choose this common denominator to be any number we please, but the simplest procedure is to choose the smallest such denominator, which, in the example given, is 15, the product of the two denominators 3 and 5. We change $1/3$ into a fraction with the denominator 15 by multiplying it by $5/5$ so that $1/3 \times 5/5 = 5/15$ and we change $2/5$ into a similar fraction by multiplying it by $3/3$ so $2/5 \times 3/3 = 6/15$. The sum of the two fractions now becomes $5/15 + 6/15 = 11/15$. Thus, $1/3 + 2/5 = (5 + 6)/(3 \times 5)$, which leads to the following general rule for adding two fractions with different denominators: Multiply the numerator of either fraction by the denominator of the other one and add these two products; this sum is the numerator of the fraction obtained by adding the two individual fractions. Now multiply the denominators of the two fractions to be added; this is the denominator of the fraction that equals the sum of the two fractions. Thus, the sum of the fraction

$$(\text{numerator}_1/\text{denominator}_1)$$

and the fraction

$$(\text{numerator}_2/\text{denominator}_2)$$

is

$$\frac{[(\text{numerator}_1)(\text{denominator}_2) + (\text{numerator}_2)(\text{denominator}_1)]}{(\text{denominator}_1)(\text{denominator}_2)}$$

The addition may be simplified if the two numerators or the

two denominators or both have common factors. Thus,

$$(4/10 + 8/25) = 4(1/10 + 2/25) = (4/5)(1/2 + 2/5)$$
$$= (4/5)(5/10 + 4/10) = 4/5 \times 9/10$$
$$= 2/5 \times 9/5 = 18/25$$

Of course we would have obtained the same result by changing the fraction $4/10$ to $4/10 \times (2\frac{1}{2}/2\frac{1}{2}) = 10/25$ so that $4/10 + 8/25 = 10/25 + 8/25 = 18/25$. Also

$$4/10 + 8/25 = (4 \times 25 + 10 \times 8)/250 = (100 + 80)/250$$
$$= 180/250 = 18/25$$

Any fraction can be changed to tenths, hundredths, etc., (to decimals) and if we do that to the two fractions to be added, the addition is like ordinary addition of integers, with the decimal points of the two decimals kept in their proper positions. Thus, $1/4 + 1/8 = 1.00/4 + 1.000/8 = 0.250 + 0.125 = 0.375$. This sum is the same as $(1/4)(1 + 1/2) = (1/4)(3/2) = 3/8$.

When any fraction is changed to a decimal the digits in the decimal either end after a definite number or they repeat themselves in identical groups (cycles). The fractions $1/2$, $1/3$, $1/4$, $1/5$, $1/6$, $1/7$, $1/8$, $1/9$ are given by the decimals $0.5000$, $0.3333 \ldots$, $0.2500$, $0.2000$, $0.16666\ldots$, $0.142857142857\ldots$, $0.125000$, $0.1111\ldots$, which we obtain by dividing the denominators into the numerator $1.0000$ as often as we please, but there is no need to continue after the digits in the decimal repeat themselves singly or in groups as in the decimal for $1/7$.

We have already discussed the multiplication of fractions but it is worth repeating the rule that the product of any number of fractions is a fraction whose numerator is the product of all the individual numerators and whose denominator is the product of all the individual denominators. Thus,

$$2/3 \times 3/5 \times 5/6 = (2 \times 3 \times 5)/(3 \times 5 \times 6)$$

$$= 30/90 = 3/9 = 1/3$$

which we could have obtained directly by crossing out common factors in the numerators and denominators, viz., $2/3 \times 3/5 \times 5/6 = 2/6 = 1/3$.

The division of one fraction by another is governed by the same rule that governs the division of integers which we can represent as a fraction by putting the dividend in the numerator and the divisor in the denominator, as in the example $6/2$ when we divide by 2. But we can also write this expression as the product of $6 \times 1/2$. Dividing by a number is thus equivalent to multiplying by its reciprocal. This rule also applies to fractions because dividing one fraction by another is the same as multiplying the first fraction by the reciprocal of the other. As an example we consider dividing $2/7$ by $2/3$; we multiply $2/7$ by $3/2$, which is the reciprocal of $2/3$. Thus, $2/7 \div 2/3 = 2/7 \times 3/2 = 3/7$. Note that when we divide any number by a fraction that is smaller than 1 (the numerator of the fraction is smaller than its denominator) the quotient so obtained is always larger than the dividend (the number being divided).

Powers applied to fractions are governed by the same rules as when powers are applied to integers. Thus, $(2/3)^3 = 2/3 \times 2/3 \times 2/3 = 2^3/3^3 = 8/27$. Note that when an integer power is applied to a fraction smaller than 1, the fraction thus obtained is always smaller than the original fraction. Thus, $(2/3)^3 < 2/3$ since $8/27 < 2/3$ (here the symbol $<$ means "less than" and $>$ means "greater than").

A negative power applied to a fraction means that we must take the reciprocal of that fraction and raise it to the same power as in the examples $(2/3)^{-1} = 3/2$, $(3/5)^{-2}$ $= (5/3)^2 = (5/3)(5/3)$. If a minus sign is applied to a fraction, it may either be applied to the numerator or to the denominator, but not to both of them. Thus,

$$-[(3 + 7)/(5 - 2)] = -(3 + 7)/(5 - 2) = (3 + 7)/-(5 - 2)$$
$$= 10/-(5 - 2) = 10/ - 3 = -3.3333$$
$$= -3\tfrac{1}{3}$$

A number written as an integer immediately followed by a fraction such as $5\tfrac{1}{2}$ means the sum of the integer and the fraction, not the product. Thus, $5\tfrac{1}{2} = 5 + 1/2 = 10/2 + 1/2$ $= 11/2 = 5.5$.

Just as integers can be used as powers of numbers, so too, can fractions, but the results are not quite the same; we have to consider the effects on the numerator and denominator of the fractional power separately. The following simple example reveals all we have to know about this arithmetic operation. We consider the number $4 = (4)^1$ $= (4)^{2/2}$ which we have simply written as the number itself raised to the power 1. But $4^{2/2} = (4^{1/2})^2$ where we have replaced the fraction $2/2$ by its equivalent $1/2 \times 2$. We know the meaning of the power 2, but what is the meaning of the power $1/2$ to which 4 is raised? Since $4 = (4)^{2/2} = (4^{1/2})^2$ $= (4)^{1/2} \times (4)^{1/2}$, $4^{1/2}$ must be that number which, when multiplied by itself (when squared), equals 4. But this is just the number 2, which is the square root of 4. Thus, $4^{1/2} = \sqrt{4} = 2$, which is a special case of the following general rule: If the exponent to which a number is raised is a fraction, the numerator of this fraction is the integer power to which the number is raised and the denominator is the root of the number. The power exponent $2/3$ means square the number and then take its cube root or take its cube root first and then square it as in the example $8^{2/3} = (8^2)^{1/3} = 64^{1/3} = \sqrt[3]{64} = 4$

$= (8^{1/3})^2 = (\sqrt[3]{8})^2 = 2^2 = 4$. Note that the root of any fraction smaller than 1 is larger than the fraction so that, as an example, $\sqrt{1/4} = \sqrt{1}/\sqrt{4} = 1/2 > 1/4$.

We now consider some other interesting properties of fractions that help us understand them better. We introduced fractions by considering points on our line that are not labeled by integers, and we did this for the points in the interval between 0 and 1. Clearly an infinite number of such points lie in that interval because no matter how close to 0 we begin in this process of assigning fractions to points we can get still closer. No matter how finely we subdivide the interval from 0 to 1 we can always set up a still finer subdivision. We may put it somewhat differently by stating that no matter how close to each other two fractions may be, an infinite number of fractions lies between them. Given the fractions 1/100 and 1/101, we can always find a fraction such as 1/100.1 larger than 1/101 and smaller than 1/100. The mathematicians describe this property of fractions by saying that the fractions on a line "lie everywhere dense." Clearly, the number of fractions in any interval of the line, however small, is infinite just like the integers on the entire line. This may lead one to the incorrect conclusion that the number of fractions on the line is far larger than the number of integers. But this is not correct because we can easily show that we can count the fractions, and when we do so, we obtain the same count as we do for the integers. This result is one of the peculiarities or surprises we run into when we deal with infinities, but we save that subject for later.

The Greek philosopher Zeno posed a series of famous paradoxes, one of which was known as the Dichotomy, which purported to demonstrate the impossibility of motion. His argument, which is relevant to our discussion of fractions, went along the following lines: "First, half the distance must be traversed, then half of the remaining distance, then again half of what remains, and so on. It follows that some portion of the distance to be covered always remains, and therefore motion is impossible!"[9] Mathematically, this situation may be described using an infinite geometric series: $1/2 + 1/4 + 1/8 + 1/16 + 1/32 + \cdots$. We can see that each term is half the magnitude of the previous term. However, the sum of this infinite geometric series is finite—it equals 1. The flaw in the paradox becomes apparent only with some thought: "Zeno assumed that any totality composed of an infinite number of parts must, itself, be infinite, whereas we have just seen an infinite number of elements which make up the finite totality—1."[10] In short, Zeno's paradox would hold true only if the sum of the infinite geometric series were infinite.

What does counting the fractions mean? How does one do it? Counting means assigning the integers 1, 2, 3, etc., in sequence to the objects. Can we do this with the fractions? In other words, can we set up a scheme which enables us to arrange the fractions in a sequence which begins with a fraction we call number 1 and goes on forever in such a way that every fraction can be placed in this sequence with a single different integer assigned to it?

We return to this idea in a later chapter and show that a number of schemes have been invented for the counting of fractions. This being so, we conclude that when we take all the integers and fractions into account on the line, we leave out most of the points on the line. Indeed, if we were to pick out a point on the line at random (with our eyes closed) the chance of picking an integer or a fraction would be zero because the number of points on the line that are not fractions or integers is infinitely larger than the countable points (integers and fractions together).

What is the nature of these numbers? To begin with, taken as a group, they are called the irrational numbers because such a number cannot be written as a fraction or as a repeating decimal. Second, they cannot be counted as already discussed above. Third, every irrational number can be approximated to any desired degree of accuracy by rational numbers (by fractions or decimals). Fourth, the most interesting numbers in mathematics and in the sciences such as $e$ and $\pi$ and the ratios of the values of certain physical quantities that describe the universe are irrational. Fifth, any irrational number can be expressed as an infinite sum of rational numbers. A simple example of an irrational number is $\sqrt{2}$, the square root of 2 which can be approximated by the number 1.4142136 ... (by a nonrepeating decimal). In the chapter on geometry we shall see how we can assign a very definite point to $\sqrt{2}$ on the line between the integers 1 and 2 or between the fractions 1.414 and 1.415.

Having introduced the irrational numbers, we return now briefly to the discussion of the logarithm of a number that lies between the powers of 10 whose logarithms are the integers. We saw that the integers 0, 1, 2, 3, 4, etc., are the logarithms of 1, 10, 100, 1000, 10,000, etc. What then is the logarithm of a number like 2 which lies between 1 and 10? It must lie between 0 and 1 (the logarithms of 1 and 10) but it cannot be expressed as a rational number (as a fraction or a repeating decimal). It is actually an irrational number which can be approximated by the decimal 0.30103 .... With this method we can find the logarithms of 20, 200, etc.; they are 1.30103, 2.30103, etc. All we need to find the logarithm of any number is a table of logarithms for the numbers between 1 and 10, which we can only approximate. We then add the integers 1, 2, 3, etc. to give us the logarithms of numbers that are 10 times, 100 times, 1000 times, etc., of the numbers between 1 and 10. Knowing the logarithms of a few numbers such as 2 and 3 enables us to deduce the logarithms of other numbers quite easily. Since $6 = 2 \times 3$, $\log 6 = \log 2 + \log 3$. Also $\log 8 = \log 2^3 = 3 \log 2$ and $\log 4 = \log 2^2 = 2 \log 2$; $\log 10 = 1 = \log (5 \times 2) = \log 5 + \log 2$. Thus $\log 5 = 1 - \log 2$. From the logarithm of 3 and the logarithm of 5 we can calculate the logarithm of 15 and so on.

## NOTES

[1] Morris Kline, *Mathematics in Western Culture*. New York: Oxford University Press, p. 41.

[2] Edward Kasner and James R. Newman, *Mathematics and the Imagination*. New York: Simon & Schuster, 1940, p. 23.

[3] *Ibid*, p. 23.

[4] James R. Newman, *The World of Mathematics*. New York: Simon & Schuster, 1956, p. 418.

[5] Carl B. Boyer, *A History of Mathematics*. Princeton: Princeton University Press, 1985, p. 111.

[6] Newman, *op. cit.*, p. 2067.

[7] Kasner and Newman, *op. cit.*, p. 80.

[8] *Ibid.*, p. 87.

[9] *Ibid.*, pp. 37–38.

[10] *Ibid.*, p. 38.

# CHAPTER 2

# Irrational Numbers, Imaginary Numbers, and Other Curiosities

*Round numbers are always false.*
—SAMUEL JOHNSON

We devoted most of the first chapter to a discussion of the rules of arithmetic and their application to integers and fractions and we used the points on a line to help us understand the rules of arithmetic. We now discuss some features of arithmetic operations that will lead us to the irrational numbers and to complex numbers. Before we proceed any further, however, we must consider special ways of representing numbers that will greatly enlarge our understanding of arithmetic. To this end we introduce first the binomial expansion which was discovered by the British scientist Isaac Newton.

We consider the number 1 which we write as the sum $1/2 + 1/2$, and we raise this sum to the successive powers 0, 1, 2, etc. Since 1 raised to any power is 1, the result is always 1 so that we obtain $(1/2 + 1/2)^0 = 1$, $(1/2 + 1/2)^1 = 1/2 + 1/2 = 1$,

$$(\tfrac{1}{2} + \tfrac{1}{2})^2 = (\tfrac{1}{2} + \tfrac{1}{2})(\tfrac{1}{2} + \tfrac{1}{2}) = \tfrac{1}{4} + 2(\tfrac{1}{4}) + \tfrac{1}{4} = 1,$$

$$\left(\tfrac{1}{2} + \tfrac{1}{2}\right)^3 = \left(\tfrac{1}{2} + \tfrac{1}{2}\right)\left(\tfrac{1}{2} + \tfrac{1}{2}\right)^2 = \tfrac{1}{8} + \tfrac{3}{8} + \tfrac{3}{8} + \tfrac{1}{8} = 1,$$

$$\left(\tfrac{1}{2} + \tfrac{1}{2}\right)^4 = \left(\tfrac{1}{2} + \tfrac{1}{2}\right)\left(\tfrac{1}{2} + \tfrac{1}{2}\right)^3 = \tfrac{1}{16} + \tfrac{4}{16} + \tfrac{6}{16} + \tfrac{4}{16} + \tfrac{1}{16} = 1,$$

etc. Each power may be expressed as a series or an expansion of a finite number of terms, with each successive series (each successive power) containing one term more than the preceding power. The denominator of the fractions in each series (expansion) is the same; the denominator in each fraction is 2 raised to the power of the expansion so that as we go from power to power (from series to series) the denominators increase by the factor 2. The sum of the numerators in any series equals 2 raised to the power of that series (expansion). In the sixth series which stems from $(1/2 + 1/2)$ raised to the fifth power, the sum of the numerators is 32. This quantity may be expressed as $2^5$ or $2 \times 2 \times 2 \times 2 \times 2$. These various series of numbers illustrate the binomial theorem which is so named because it deals with a sum of two terms (in this example, $1/2 + 1/2$) raised to various powers. This theorem has important applications in mathematics and physics.

We point out a few more interesting features of the binomial expansion which illustrate its usefulness and begin with the numerators of the fractions in each series. These numbers, known as the binomial coefficients, play an important role in the laws of probability, as we shall show presently. We see that in each series these coefficients begin with 1; then increase to a maximum value and then decrease through the same numbers through which they increased so that the coefficients in the second half of the expansion are the same as those in the first half. The second coefficient in each expansion is exactly the power of the expansion and the third coefficient is the power multiplied by the power minus 1 divided by 2, and so on. We illustrate this operation with the expansion of $(1/2 + 1/2)^5$. The first coefficient is 1, the second is 5, the third is $5(5 - 1)/2$, the fourth is $5(5 - 1) \times (5 - 2)/(2 \times 3)$, and so on.

If we arrange all the coefficients in all the expansions in separate lines (one line for each power) to form a triangle, we obtain the famous Pascal triangle which was discovered by the brilliant French mathematician Blaise Pascal:

|   |   |   |   |   |                |
|---|---|---|---|---|----------------|
|   |   | 1 |   |   | zero power     |
|   | 1 |   | 1 |   | first power    |
| 1 |   | 2 |   | 1 | second power   |
|1  | 3 |   | 3 |   |1  third power  |
|1 | 4 | 6 | 4 | 1 | fourth power   |

etc. This triangle shows the beautiful symmetry of the binomial coefficients; each succeeding line (the base) of the triangle is obtained from the line above it by adding each number in that line to its left and right neighbors. Note that the sum of the coefficients in any line equals 2 raised to the power represented in that line. The third line, for example, represents $2^3 = 8 = 1 + 3 + 3 + 1$. That these coefficients play an important role in probability theory is best illustrated by the "coin tossing" problem which may be stated as follows: Given a certain number of coins which are tossed into the air simultaneously, what is the chance (the probability) that a given fraction of them will come up heads and the rest tails, such as, for example, all heads or all tails, or 1/3 heads and 2/3 tails?

To see how the binomial coefficients answer this question we must first define precisely the probability for a certain event to occur if a group of such events is possible. This probability is defined as the number of favorable events (the events we want to happen) as a fraction of the total number of events. If we have 7 white marbles and 5 red ones mixed together in a bag and pick a marble out of the bag at random, the chance (probability) that the marble is red is 5/12 and the chance that it is white is 7/12. Note that the sum of the two probabilities is 1, which means that we will certainly have chosen some marble. In probability theory 1 means a certainty so that any event that is not certain to happen must have a probability less than 1. Probabilities are always expressed as percentages which, as we have seen, are the same as fractions. Thus, a probability of 25% is the same as 1/4.

With this definition of probability understood, we return now to the "coin tossing" problem. Considering a single coin, we see that only two events are possible: either the head or the tail of the coin will come up when it is tossed so that each has a probability of 1/2. We may represent the tossing of a single coin by the sum $(1/2 + 1/2)$, the first 1/2 representing the chance that a head (or a tail) will come up and the second 1/2 the chance that a tail (or a head) will come up. Since either a head or a tail must come up, the sum of the two probabilities must be 1 which is given by $1/2 + 1/2$.

We now consider two identical coins that are tossed simultaneously and calculate the probability of each of the various possible combinations of heads and tails (two heads, two tails, one head and one tail). Since the two coins are independent of each other, the probability for the coins to come up in any particular combination of heads and/or tails is given by the product of the two individual probabilities, that is, $(1/2 + 1/2)(1/2 + 1/2)$ or $(1/2 + 1/2)^2$. The probabilities for all the possible combinations of heads and tails are given by the coefficients in this binomial expansion. These coefficients are 1, 2, 1; the probabilities are given by dividing each of these numbers by 4, which gives us 1/4, 1/2, 1/4. The chance for either two heads or two tails is thus 1/4 (25%)

and the chance for a head and a tail is 1/2 (50%). If we toss five coins, the probabilities are given by the terms in the binomial expansion of $(1/2 + 1/2)^5$, which gives us 32 terms so that 32 is in the denominator of each term in the expansion as shown above. The probability for all five heads or all five tails is thus 1/32; the probability for four heads and one tail or one head and four tails is 5/32, and so on.

The value of the discovery of the laws of probability cannot be overestimated for they have applications in all areas of human knowledge ranging from mathematics and the sciences to philosophy and psychology. According to Eric Temple Bell, Pascal's theory of probabilities brought "the superficial lawlessness of pure chance under the domination of law, order, and regularity, and today this subtle theory appears to be at the very roots of human knowledge no less than at the foundation of physical science."[1] This intellectual achievement was representative of the efforts begun by a number of individuals in 17th century Europe to analyze nature in a more rigorous and logical fashion. Doubtless Pascal's work gave some impetus to Isaac Newton's formulation of his three laws of motion and his law of universal gravitation in the 1660s; these laws formed the basis for Newton's mechanistic universe in which the cosmos was thought to be governed by certain immutable physical laws. Newton believed that the universe had been created and set in motion by God but that it, like a watchspring, had been left to run down on its own accord.

Pascal and Pierre de Fermat developed the theory of probability during an intensive correspondence they carried out in the year 1654. Their letters reveal an interesting collaboration in which each man posed problems which they solved using certain basic principles of probability which they developed together. What distinguished the work of each man

was not the basic principles (about which they were in fundamental agreement) but the much less important details of the
solutions themselves. In any event, the problem which led
Pascal to begin his search for a theory of probability was
posed to him by a professional gambler named Chevalier de
Méré. Chevalier wanted to know how the stakes in a particular
game should be divided when a certain number of points
were required to win the game but the players chose to quit
the game before they finished it: "The score (number of
points) of each player is given at the time of quitting, and
the problem amounts to determining the probability which
each player has at a given stage of the game of winning the
game. It is assumed that the players have equal chances of
winning a single point. The solution demands nothing more
than sound common sense; the mathematics of probability
enters when we seek a method for enumerating possible cases
without actually counting them off."[2]

Although the utility of Pascal's mathematical theory of
probability had only begun to be demonstrated in his own
lifetime, the contributions of other individuals made it possible for the theory to evolve over time so that it became
inextricably woven with the fabric of mathematics as a whole.
To attempt to remove probability theory from modern mathematics would be akin to removing the skeleton from a body:
The backbone of the organism would no longer exist and the
remains would be little more than a shapeless remnant of its
former self. However, Pascal did employ his theory of probability to conclude that he should lead a pious life so that he
might maximize his chance for winning the prize of eternal
happiness in heaven. Because the motivation for any wager
is the value of the prize multiplied by the probability of
winning the prize, Pascal concluded that one should lead a
pious life because the value of everlasting life in heaven is

infinite even though the possibility of receiving the prize might be remote. Because the value assigned to such a prize is infinite, the laws of probability will invariably favor such a course of behavior. This is not to say that Pascal was enthralled with passing on the more intoxicating of life's pleasures, but he did recognize that probabilistic calculations could come into play in all human endeavors.

We now use the binomial expansion to calculate the square root of 2 (which we write as $\sqrt{2}$). This can be done with the standard algorithm that one learns in elementary school but that calculation is generally done by rote without any understanding of why the algorithm works, whereas everything is less cryptic in the application of the binomial expansion. We have $\sqrt{2} = \sqrt{(1+1)} = (1+1)^{1/2}$. That the exponent in this example is not an integer but a fraction should not deter us because the expansion is valid regardless of the nature of the exponent. Following the rule for the expansion we have

$$(1+1)^{1/2} = 1^{1/2} + (1/2)1^{1/2-1}1$$

$$+ \frac{(1/2)(1/2-1)}{2} 1^{1/2-2}1^2$$

$$+ \frac{(1/2)(1/2-1)(1/2-2)}{2 \times 3} 1^{1/2-3}1^3$$

$$+ \frac{(1/2)(1/2-1)(1/2-2)(1/2-3)}{2 \times 3 \times 4} 1^{1/2-3}1^3$$

$$= 1 + 1/2 - 1/8 + 1/16 - (5/4)/32 + \cdots$$

Before continuing with our discussion of $\sqrt{2}$ as given by the binomial expansion, we explain the appearance of the exponents of 1 in the expansion. When we wrote down the successive terms in the binomial expansion initially we

explained the way the coefficients in the expansion are calculated but we said nothing about the exponents, which we now
discuss. The simple example $(2 + 3)^2$ is sufficient to illustrate
the rule for the exponents. The square of $(2 + 3)$ means that
we must multiply each term in the sum by itself and by the
other term so that we get four terms in the final product, each
of which contains $2^2$, $3^2$, or the product $2 \times 3$. In fact, we get
$2^2 + 2(2 \times 3) + 3^2$ which gives us the following rule for the
exponents: The exponent of 2 is 2 in the first term, 1 in the
second term, and 0 in the last term, whereas the exponent of
3 is 0 in the first term, 1 in the second term, and 2 in the
last term. This statement may be expressed as $(2 + 3)^2$
$= 2^2 \times 3^0 + 2(2^1 \times 3^1) + 2^0 \times 3^2$ noting that $3^0 = 1$ and $2^0 = 1$.
Thus, 2 starts with the largest possible exponent which
decreases by 1 in each successive term until it reaches 0,
whereas the exponent of 3 starts with 0 and increases by 1
in each successive term until it reaches 2. Note also that in
each term the two exponents must add up to 2. Applying this
idea to the expansion $(2 + 3)^3$ which contains eight terms
(only four terms are different) we have the following products
of the powers of 2 and 3, starting with the first term: $2^3 \times 3^0$,
$2^2 \times 3^1$, $2^1 \times 3^2$, $2^0 \times 3^3$; the power of 2 diminishes a unit at
a time and that of 3 increases in the same way. In each term
the sum of the two exponents is 3.

   With this rule for the exponents of the terms in the
binomial expansion understood, we see why the expression
for $\sqrt{2}$ has the various powers of 1 we wrote previously. Since
1 to any power equals 1, we could have left out all the powers
of 1 such as $1^{1/2 - 2}$ and replaced each of them by 1 and then
left the factor 1 out because the factor 1 is always understood.
In any case, the binomial expansion leads us to an infinite
series of rapidly diminishing terms. Since the signs of the
successive terms in the series alternate between + and −, this

series is called an alternating series, which converges to a finite value, namely $\sqrt{2}$, as the number of terms is increased. We have carried out the binomial expansion for $\sqrt{2}$ to the fifth term which still gives a fairly accurate value for $\sqrt{2}$. We can get as close to $\sqrt{2}$ as we wish by increasing the number of terms in the expansion.

The binomial expansion is particularly useful if one of the two terms in the sum to be expanded is small. As an example, we consider $\sqrt{5}$, which we write as $\sqrt{(4+1)}$ or $\sqrt{4(1+1/4)}$, where we have placed 4 as a separate factor, as shown by the parentheses. This expression may be rewritten as follows:

$$\sqrt{4(1+1/4)} = \sqrt{4}\sqrt{(1+1/4)} = 2\sqrt{(1+1/4)} = 2(1+1/4)^{1/2}$$

The binomial expansion now gives us

$$\sqrt{5} = 2\{1^{1/2} + (1/2) \times 1^{1/2-1} \times 1/4$$
$$+ (1/2)(1/2-1)/2(1^{1/2-2})(1/4)^2$$
$$+ ([(1/2)(1/2-1)(1/2-2)]/(2 \times 3)1^{1/2}$$
$$\times (1/4)^3 + \cdots\}$$

Note that the second term contains the factor $1/4$, the third term the factor $1/16$, the fourth term the factor $1/64$, and so on in the series so that these terms become very small and may be neglected. A still better example of the usefulness of the binomial expansion in obtaining a good approximation using only a few terms is calculating $\sqrt{10} = \sqrt{(9+1)} = \sqrt{9}\sqrt{(1+1/9)} = 3(1+1/9)^{1/2}$. Here the second, third, and fourth terms contain the factors $1/9$, $1/81$, and $1/729$ so that the series converges very quickly.

If we know the square roots of a few prime numbers such as 2, 3, 5, 7, 11, we can get the square roots of many more numbers by multiplication. Thus, $\sqrt{6} = \sqrt{(3 \times 2)} = \sqrt{3} \times \sqrt{2}$; $\sqrt{10} = \sqrt{(2 \times 5)} = \sqrt{2} \times \sqrt{5}$; $\sqrt{15} = \sqrt{(5 \times 3)} = \sqrt{5} \times \sqrt{3}$, and so on.

We can use the binomial expansion to obtain cube roots, fourth roots, etc. We use the same expansions as we did for the square roots, replacing 1/2 in these expansions by 1/3 for the cube root, 1/4 for the fourth root, and so on. Thus,

$$\sqrt[3]{2} = (1 + 1)^{1/3} = (1 + 1/3 + [(1/3)(1/3 - 1)]/2$$
$$+(1/3)[(1/3 - 1)(1/3 - 2)]/(2 \times 3)$$
$$+(1/3)[(1/3 - 1)(1/3 - 2)$$
$$\times(1/3 - 3)]/(2 \times 3 \times 4) + \cdots$$

We now use the binomial expansion to represent 2 by an infinite series. We have $2 = 1/(1 - 1/2) = (1 - 1/2)^{-1}$ and recall that a factor in a denominator equals the same factor in the numerator with the exponent $-1$. Since the rule for the binomial expansion with a negative exponent is the same as that for a positive exponent, we obtain

$$(1 - 1/2)^{-1} = [1 + (-1/2)]^{-1}$$
$$= 1^{-1} + (-1) \times (-1/2)$$
$$+[(-1)(-1 - 1)/2](-1/2)^2$$
$$+[(-1)(-1 - 1)(-1 - 2)/(2 \times 3)](-1/2)^3$$
$$+ \cdots$$
$$= 1 + 1/2 + 1/4 + 1/8 + \cdots$$

This is an example of a very important infinite series called the geometric series, which, in this case, is obtained by raising 1/2 to all possible powers (including 0) and summing these powers. Power series, of which the above expansion is one example, are very important and useful in physics and mathematics.

Expressions such as $\sqrt{2}$, $\sqrt{3}$, etc., and higher roots of integers are, in general, neither integers nor fractions; they are members of a much larger set of numbers called irrational numbers. But the irrational numbers include another set of numbers called transcendental numbers, the most famous and important of which are $e$ and $\pi$. The number $e$ can be obtained as an infinite series by considering the binomial expansions $(1 + 1/2)^2$, $(1 + 1/3)^3$, $(1 + 1/4)^4$, and so on, allowing the exponents of the expression and the denominators in the second terms to get larger and larger simultaneously. We get the series for $e$ when the exponent and the denominator become infinite. To see what we get we consider the expansion of $(1 + 1/n)^n$ where $n$ stands for any number we please. That $n$ instead of some specific number is used makes no difference as far as the rule for the binomial expansion is concerned. We have the following series of terms, each of which has $n$ to some positive power in its denominator:

$$\left(1 + \frac{1}{n}\right)^n = 1^n + n \times 1^{n-1} \times \left(\frac{1}{n}\right)$$

$$+ \frac{n(n-1)}{2} \times 1^{n-2} \times \left(\frac{1}{n}\right)^2$$

$$+ \frac{n(n-1)(n-2)}{2 \times 3} 1^{n-3} \left(\frac{1}{n}\right)^3 + \cdots$$

$$= 1 + 1 + \frac{1}{2} - \frac{1}{2n} + \frac{1}{2 \times 3} - \frac{3}{2 \times 3}$$

$$\times \frac{1}{n} + \frac{2}{2 \times 3} \times \frac{1}{n^2} + \cdots$$

$$= 1 + 1 + \frac{1}{2} + \frac{1}{2 \times 3} + \frac{1}{2 \times 3 \times 4} + \cdots$$

plus terms each of which has $n$ to some positive power in its denominator. If $n$ is made very large, all the terms with $n$ in their denominators become very small and may therefore be dropped. We are thus left with the series $1 + 1 + 1/2 + 1/(2 \times 3) + 1/(2 \times 3 \times 4) + \cdots$ for $e$ whose value is 2.718281828459 out to 12 decimal places. Note that the integers following the decimal do not repeat themselves in any pattern or in any cycle like the decimal form of a fraction. This means that $e$ is not a rational number unlike all fractions; nor is it the root of any integer or fraction. For that reason, $e$ is called a transcendental number, one of a set of numbers which also contains the number $\pi$. We return to $\pi$ later but we note here that it is described by an infinite series: $\pi = 4(1 - 1/3 + 1/5 - 1/7 + \cdots)$. It has been calculated to many decimal places but it will suffice for our purposes to give it to 12 such places: 3.141592653589. Here, too, integers do not repeat themselves in any cycle.

We come now to another way of representing a number which has a curious attraction for us because it has the same mystical appearance as an infinite spiral staircase that never ends but seems to lead to nowhere or an image that is repeated over and over again in appropriately placed mirrors. We are referring here to what mathematicians call continued fractions. We may illustrate this concept by expressing 2 as a continued fraction:

$$2 = \cfrac{1}{1 - \frac{1}{2}} = \cfrac{1}{1 - \cfrac{1}{1 - \cfrac{1}{1 - \frac{1}{2}}}} = \cfrac{1}{1 - \cfrac{1}{1 - \cfrac{1}{1 - \cfrac{1}{1 - \frac{1}{2}} + \cdots}}}$$

We can represent any irrational number as a continued fraction, the simplest example of which is

$$\sqrt{2} = 1 + \cfrac{1}{2 + \cfrac{1}{2 + \cfrac{1}{2 + \cfrac{1}{2 + \cfrac{1}{2} + \cdots}}}}$$

We can easily verify that this is a correct equation by writing it first as

$$\sqrt{2} = 1 + \cfrac{1}{1 + 1 + \cfrac{1}{2 + \cfrac{1}{2 + \cfrac{1}{2} + \cdots}}}$$

and noting that the sum of terms after the 1 in the first denominator is the same as the entire continued fraction; it equals $\sqrt{2}$ according to our initial assumption. We thus have $\sqrt{2} = 1 + 1/(1 + \sqrt{2})$. Since this is an equation, we may do anything arithmetic to the left side if we also do the same to the right side; we must treat the two sides equally. We may therefore multiply both sides by $(1 + \sqrt{2})$. We thus have

$\sqrt{2}(1 + \sqrt{2}) = 1 + \sqrt{2} + (1 + \sqrt{2})/(1 + \sqrt{2})$ or $\sqrt{2} + 2 = 1 + \sqrt{2}$ $+ 1$. We thus see that the continued fraction for $\sqrt{2}$ is correct because it leads to the identity $\sqrt{2} + 2 = \sqrt{2} + 2$, which shows that the left side $\sqrt{2}$ of our initial equation, equals the right side of the continued equation.

Before leaving this discussion of irrational numbers we note an important difference between these numbers as a set or a class and the rational numbers which consist of all the integers and fractions. Since both the rational and irrational numbers locate points on a line once we have chosen the point 0, and all points on a line are identical, there is no difference between these two sets or classes of numbers, insofar as the nature of the associated points on a line is concerned. The two sets of numbers differ greatly, however, in the quantity of points they locate; the number of points to which the irrational numbers are attached is infinitely larger than the number of points to which the rational numbers are assigned.

We may express this idea somewhat differently in a way which involves the concept of countability—the idea that a set of numbers or other entities can be counted. To pursue this thought we must first define carefully the counting operation; this operation means devising a scheme which permits us to associate each entity in one set with an integer— that is, setting up a one-to-one correspondence between the entities and the integers. There is no difficulty with this operation if the number of entities in our set is finite, but if the number is infinite some difficulties and ambiguities arise. Of course, we have no difficulty dealing with the integers themselves because they count themselves, by the very definition of counting. But some difficulty arises in trying to comprehend the statement, which we shall prove, that the fractions are countable or, as the mathematician states it, denumerable. The reason for this difficulty is that when we assign the integers to points on a line, they are spread out and no integer lies between any two successive integers. When we assign fractions to the points between those labeled by integers, the fractions appear to be infinitely crowded together; no matter how close to each other two points labeled by fractions are located, there is still room between them for an infinite number of other such points (other fractions). That the fractions are so closely crowded together, compared to the integers, seems to contradict our statement that the fractions can be counted and that they are therefore no more numerous than the integers. But this apparent contradiction is not real as we shall now show.

We include in our first set of fractions to be counted only the "proper fractions" defined as those fractions whose denominators are larger than their numerators (fractions $< 1$); we exclude the "improper fractions" such as $3/2$, $1/1$, $2/2$, etc. We now describe a simple scheme for counting all the fractions which permits us to assign an increasing integer starting with 1 to every fraction in turn in an orderly way that does not overlook a single fraction and is such that given any fraction we know at once which integer must be assigned to it. We proceed by adding the numerator and denominator of a given fraction; the smaller the sum, the lower is the integer (the count) we assign to the fraction. If this sum is the same for two different fractions, we assign the lower count (smaller integer) to the fraction with the smaller numerator. We illustrate this counting scheme by arranging twelve fractions in a sequence:

| 1/2 | 1/3 | 1/4 | 2/3 | 1/5 | 2/4 | 1/6 | 2/5 | 3/4 | 1/7 | 2/6 | 3/15 |
|-----|-----|-----|-----|-----|-----|-----|-----|-----|-----|-----|------|
| 1   | 2   | 3   | 4   | 5   | 6   | 7   | 8   | 9   | 10  | 11  | 12   |

We see from this small sample that every fraction will ultimately have an integer assigned to it. Note that for counting purposes we treat 1/2 and 2/4 as different fractions even though they have equal values.

We now show that the irrational numbers cannot be counted, for no matter how we arrange them in a sequence we always leave infinitely many numbers out. Let us consider only those irrational numbers that are <1 and are approximated by decimals that have no repeating cycles of integers. If we arrange them in a column, one irrational number under the other such as

1. 0.2574951 ...
2. 0.3687924 ...
3. 0.1469432
4. 0.6578643

and so on, numbering each one, starting with the integer 1 and continuing onward indefinitely, we cannot possibly include all the irrational numbers in any such columnar array because we can always construct or find an irrational number that is not contained in the array. We simply construct an irrational number whose first integer differs from the first integer of the top irrational number, and whose second, third, fourth, etc. integers differ from the second, third, fourth, etc., integers in the second, third, fourth, etc., irrational numbers. Thus, we construct the irrational number 0.4789 .... This number is certainly not included in the array given above because it differs in at least one integer from every number listed above.

We complete this chapter with the definition and discussion of complex numbers which were considered as mere curiosities in the early history of mathematics but which form an important branch of mathematics today and play a very important and useful role in almost all branches of science, particularly in physics, astronomy, and chemistry. The theory of complex numbers and the vast branch of mathematics called complex variables grew out of the number $\sqrt{-1}$ which is called the imaginary unit and labeled $i$. It is called imaginary because it cannot be assigned to any point on a line; nor does any number (rational or irrational) exist whose square is $-1$. If we square any positive or negative real number, we always obtain a positive number so that $\sqrt{-1}$ or $i$ lies in a whole new domain of numbers called imaginary numbers which cannot be obtained from real numbers by any numerical operation.

Having introduced the letter $i$ for $\sqrt{-1}$, the imaginary unit, we see that it has the same validity as a number as the other ten symbols, 0, 1, 2, etc. We can change any real number such as 2 into an imaginary number by multiplying it by $i$. Thus, $i4$ (or $4i$) is a pure imaginary number. If we now multiply $4i$ by $i$ we can obtain $4i \times i = 4i^2 = -4$, the negative of 4. From this discussion we see that we can change a pure real number into a pure imaginary number by multiplying the real number by $i$; multiplying a pure imaginary number by $i$ gives a pure real number with the opposite sign.

Since $i = \sqrt{-1}$ or $(-1)^{1/2}$, we obtain the following equalities: $i^2 = i \times i = (\sqrt{-1})^2 = -1$, $i^3 = i \times i^2 = i \times (-1) = -i$, and $i^4 = i^2 \times i^2 = (-1)(-1) = 1$. We thus get back to 1 by multiplying $i$ by itself three times. Thus, the four quantities 1, $i$, $i^2$, $i^3$ or 1, $i$, $-1$, $-i$ form a small self-contained group such that multiplying any number of this group by itself or by any other member gives a member of the group. Another interesting property of $i$ is that its reciprocal equals its own negative: $1/i = -i$. That this is a correct equation can be seen if we multiply both of its sides by $i$ which does not invalidate the equation. We then have $i \times 1/i = -i \times i$ or $1 = -i^2 = 1$.

Since $i$, as a number, has all the rights and privileges of any real number, we may do with $i$ anything we are permitted to do with a real number. We may thus raise any real number to the power $i$, for example, $4^i$, $5^i$, and even $i^i$, and we may take the roots of $i$ such as $\sqrt{i} = i^{1/2}$, $i^{1/3}$, etc. We shall see later just what these quantities mean.

We have thus far considered the real and imaginary numbers separately, but no arithmetic rule forbids us to add an imaginary number to a real number. We then obtain a complex number such as $2 + 5i$, which we must leave in that form since the real number 2 and the imaginary number $5i$ cannot be combined into a single entity.

Having introduced or defined a complex number as a sum of a real number and an imaginary number, we can now operate with complex numbers just as we do with real numbers; all the arithmetic rules that apply to real numbers, apply without restriction to complex numbers. We may add them, divide them, raise them to powers, take their roots, and obtain their logarithms.

Before we illustrate these operations using specific complex numbers, we introduce some additional concepts that simplify these operations. The first of these additional concepts is the complex conjugate of a complex number, which we designate by an asterisk placed in the upper right-hand side of the complex number so that $(4 + 2i)^*$ means the complex conjugate of $4 + 2i$. The complex conjugate of any complex number is obtained by replacing $i$ wherever it may appear by $-i$; note that this rule applies to all $i$'s no matter how often they appear in the complex number. In the simple example $4 + 2i$, we have $(4 + 2i)^* = 4 - 2i$, but when we have a number such as

$$\left(\frac{4 - 2i}{3 + 7i}\right)\sqrt{(1 + 3i)}$$

we would find it quite difficult to write down its complex

conjugate, if we did not have the rule stated above. Thus,

$$\left[\left(\frac{4-2i}{3+7i}\right)\sqrt{(1+3i)}\right]^* = \left(\frac{4+2i}{3-7i}\right)\sqrt{(1-3i)}$$

Note that the complex conjugate of a product of complex numbers is the product of the complex conjugates of the number. Thus,

$$\left[\left(\frac{4-2i}{3+7i}\right)\sqrt{(1+3i)}\right]^* = \frac{(4-2i)^*}{(3+7i)^*}\sqrt{(1+3i)}^*$$

Consider now a complex number such as $4+2i$ and its complex conjugate $4-2i$. If we add these numbers together, we obtain $2 \times 4$ and if we subtract the complex conjugate from the complex number we obtain $2(2i)$. This means that if we add to a complex number its complex conjugate (complex number + complex number*), we obtain twice the real part of the complex number. If we subtract from a complex number its complex conjugate, we obtain twice the imaginary part of the complex number. We further note that if the complex conjugate of a number equals the number, the number must be real, but if the complex conjugate equals the negative of the number, the number is a pure imaginary one. Under ordinary circumstances we can determine whether a number is real, pure imaginary, or complex. The three numbers $4$, $4+2i$, and $2i$ are real, complex, and pure imaginary numbers, respectively. What about the number $i^i$ which is the number $i$ raised to the $i$th power? We apply our complex conjugate rule by replacing $i$ by $-i$ in this number which reveals its nature immediately. We have $(i^i)^*$ $= (-i)^{-i} = (1/i)^{-i} = i^i$ so that $i^i$ is real because its complex conjugate equals itself.

We now discuss the addition and multiplication of complex numbers; these operations are governed by the same rules that govern these arithmetic operations applied to real numbers. When adding complex numbers together, however, the real and imaginary parts of the numbers are kept separate. In short, the real parts of the complex numbers are added to each other as are the imaginary parts of the complex numbers. As an example we have $(4 + 2i) + (5 + 6i) = (5 + 4) + (2i + 6i) = 9 + 8i$. When multiplying two complex numbers, however, we apply the distributive rule and multiply each term of one of the complex numbers with each term of the other. Thus,

$$(4 + 2i) \times (5 + 6i) = 4 \times 5 + 4 \times 6i + 5 \times 2i + 2i \times 6i$$
$$= 20 + 24i + 10i + 12 \times i \times i$$
$$= 20 + 34i + (12) \times (-1)$$
$$= (20 - 12) + 34i = 8 + 34i.$$

An important property of a complex number is called its absolute value, which may be written, for example, as $|4 + 2i|$. The absolute value is defined as the square root of the square of the real part of the complex number minus the square of its imaginary part. Thus, the absolute value of the complex number $4 + 2i$ is $[4 \times 4 - (2i) \times (2i)]^{1/2} = (16 - 4i^2)^{1/2} = (16 + 4)^{1/2} = \sqrt{20}$. Note that the absolute value of any complex number is also the square root of the product of the number and its complex conjugate. In the example above

$$|4 + 2i| = [(4 + 2i)(4 + 2i)^*]^{1/2} = \sqrt{(4 + 2i)(4 - 2i)}$$
$$= \sqrt{16 + 4} = \sqrt{20} = \sqrt{4 \times 5} = 2\sqrt{5}$$

This definition of the absolute value of a complex number is important because we often meet complex numbers whose real and imaginary parts we cannot easily discern. Thus, the real and imaginary parts of the complex number $\sqrt{(4+2i)}$ are not evident, but its absolute value is the square root of the product $\sqrt{(4+2i)} \times \sqrt{(4+2i)}^* = \sqrt{(4+2i)} \times \sqrt{(4-2i)} = \sqrt{(16+4)} = \sqrt{20}$, that is, $|\sqrt{(4+2i)}| = \sqrt{\sqrt{20}} = (20)^{1/4}$. We can write $\sqrt{(4+2i)}$ as the sum of a real and pure imaginary part by using the rule that the real part of a complex number is $1/2$ the sum of the number and its complex conjugate so that the real part of $\sqrt{(4+2i)}$ is $(1/2)[\sqrt{(4+2i)} + \sqrt{(4-2i)}]$. The imaginary part is $1/2$ the difference between the number and its complex conjugate. Thus, the imaginary part of $\sqrt{(4+2i)}$ is $(1/2)[\sqrt{(4+2i)} - \sqrt{(4-2i)}]$. If we now square the real part we obtain $(\tfrac{1}{2})^2[\sqrt{(4+2i)} + \sqrt{(4-2i)}]^2$ which is just $(\tfrac{1}{4})[4 + 2i + 2\sqrt{(4+2i)}\sqrt{(4-2i)} + 4 - 2i]$ $= (\tfrac{1}{4})(8 + 2\sqrt{20}) = 2 + \tfrac{1}{2}\sqrt{20}$. This is the square of the real part so that the real part is $[2 + (1/2)\sqrt{20}]^{1/2}$. We now square the imaginary part and obtain

$$(\tfrac{1}{2})^2[\sqrt{(4+2i)} - \sqrt{(4-2i)}]^2$$
$$= \tfrac{1}{4}\{4 + 2i - 2\sqrt{(4+2i)(4-2i)} + 4 - 2i\}$$
$$= \tfrac{1}{4}(8 - 2\sqrt{20}) = (2 - \tfrac{1}{2}\sqrt{20})$$

The square root of this number is the imaginary part of $\sqrt{4+2i}$. We can show it as imaginary by writing it as

$$\sqrt{-(-2 + \tfrac{1}{2}\sqrt{20})} = \sqrt{(-1)(-2 + \tfrac{1}{2}\sqrt{20})}$$
$$= \sqrt{-1}\,\sqrt{2 + \tfrac{1}{2}\sqrt{20}}$$
$$= i\sqrt{-2 + \tfrac{1}{2}\sqrt{20}}$$

Thus,

$$\sqrt{(4+2i)} = \sqrt{2 + \tfrac{1}{2}\sqrt{20}} + i\sqrt{-2 + \tfrac{1}{2}\sqrt{20}}$$

is written as the sum of a real and an imaginary part and we see, using the standard definition, that the absolute value of $\sqrt{4+2i} = (\sqrt{20})^{1/2}$.

We complete our discussion of complex numbers by describing the graphical method of representing complex numbers which was developed in the mid-19th century by the great German mathematician Carl Friedrich Gauss. We have seen that we can represent all the real numbers by points on a line. We start with such a horizontal line and through the point on it labeled 0 we draw a vertical line perpendicular to the given line as Gauss did. Following Gauss we represent all the pure imaginary numbers as points on the vertical line above the line of real points and negative imaginary numbers as points on the vertical line below the line of real numbers. Thus, the point one unit above 0 is labeled $i$ and the point two units above 0 is labeled $2i$, and so on. The point one unit below 0 is $-i$ and so on. The horizontal line is called the real axis and the vertical line is called the imaginary axis.

The points on these two intersecting lines give us only the real numbers and the pure imaginary numbers. To obtain a complex number we must leave these lines and move to the points in the plane between the lines. This plane is called the complex plane since every point on it defines a complex number. Thus, the complex number $4 + 2i$ is the point on the plane we reach if we move four units to the right of 0 along the real axis and then move vertically upward two units. The absolute value of a complex number (any point on the plane) is the length of the line from 0 to the point that represents the complex number. All the properties of complex numbers can be easily derived from their representation as points on the complex plane.

## NOTES

[1] Eric Temple Bell, *Men of Mathematics*. New York: Simon & Schuster, 1937, p. 86.
[2] *Ibid.*, pp. 86–87.

# CHAPTER 3

# From Arithmetic to Algebra

*Algebra begins with the unknown and ends with the unknowable.*
—ANONYMOUS

In our relationship to numbers in our daily lives we are interested in specific numbers and not just numbers in a general or abstract way. Thus, each of us wants to know such things as our blood pressure, bank balance, and Social Security number, but as long as we limit ourselves to specific numbers, the usefulness of arithmetic itself is limited. If we write the symbol "5" we think of it as representing five and only five things, but since the laws of arithmetic apply to any number of things, we extend our arithmetic by operating not with specific numbers but with entities that may represent any number or numbers. This extension of our arithmetic to include symbols that may stand for or represent any numbers is called algebra. As we shall see, the rules of algebra are identical to those of arithmetic, but they apply to numbers combined with symbols rather than to numbers alone. Since the symbols in algebra stand for numbers, even though the numbers are not specific, algebra is the arithmetic of symbols.

Mathematics may be regarded as an intellectual endeavor in which the mathematician begins with certain concepts and deduces particular conclusions from those concepts by drawing upon an array of rules which provide guidance as to how the reasoning process is to be carried out. According to Morris Kline, "reasoning about numbers—if one is to go beyond the simplest procedures of arithmetic—requires the mastery of two faculties, vocabulary and technique, or one might say, vocabulary and grammar."[1] In addition, "the entire language of mathematics is characterized by the extensive use of symbolism [and] it is the use of symbols and of reasoning in terms of symbols which is generally regarded as marking the transition from arithmetic to algebra, although there is no sharp dividing line."[2]

The use of alphabetic instead of numerical symbols in algebra causes more than a few students to throw up their hands in dismay. Contrary to popular belief, however, these symbols are used not to torture students but instead to facilitate the ease with which algebraic equations can be expressed and understood. It is not the aim of mathematicians to impress nonspecialists with vast arrays of symbols arranged in equations that could have been found on one of the levels of Dante's *Inferno* but, instead, to provide a distinct language which is both precise and compact. By using certain algebraic symbols, the mathematician can "write lengthy expressions in a compact form so that the eye can see quickly and the mind can retain what is being said."[3] A few symbols can express a relationship that cannot be described in less than a full page of prose. Moreover, symbols with precise meanings bring a certainty to mathematical reasoning that is not available in the social sciences, for example, because the symbols have been purged of any inherent ambiguities. "By using symbols for specific ideas mathematicians avoid ambiguity or, to put the matter positively, each symbol has its own precise meaning, and so the resulting expressions are clear."[4]

This does not mean that the branch of mathematics called algebra has always been characterized by symbols; the ancient civilizations in Egypt, Babylonia, and Greece performed many algebraic operations with words. This so-called rhetorical algebra involved "a process of algebraical reasoning that did not use any symbols."[5] However, its development was not even and "the results attained as the net outcome of the work of six centuries on the theory of numbers [in the ancient world] are, whether we look at the form or the substance, unimportant or even childish, and are not in any way the commencement of a science."[6] Only after science began to reach a certain level of complexity in post-Renaissance Europe did scientists and mathematicians begin to abbreviate algebraic operations with symbols to avoid unnecessarily time-consuming calculations.

Returning now to the representation of numbers by points on a line, we put the symbol $x$ under the line to represent any point whether it is marked by an integer, a fraction, or an irrational number. Just as we became accustomed to working with the ten symbols $0, 1, 2, \ldots, 9$ to express specific numbers, so we quickly become accustomed to operating with the symbol $x$, which represents any number, keeping in mind that the basic rules stated and illustrated in the first two chapters for specific numbers apply to symbols such as $x$, which may represent any number. Algebra is, then, the arithmetic of symbols that are used to represent numbers, real, imaginary, or complex.

In using $x$ to designate points on a line we are assigning an ordinal property to $x$ and so we can use $x$ to represent an ordered array of entities whether they are points on a line or, for example, houses on a street. In labeling the points on a line with $x$, we still start from some point 0 as the origin and use $-x$ to represent points to the left of 0 if $x$ represents points to the right of 0. We also introduce a unit of length which enables us to find the point represented by $x$ by laying off $x$ units to the right of 0, or $x$ units to the left of 0 for the point $-x$. If $x$ represents some particular point, then $3x$ represents a point 3 times farther to the right of 0 than the point $x$ and $(1/2)x$ or $x/2$ or $0.5x$ represents a point one-half as far to the right of 0 as is the point $x$. If we let $a$ represent any number (not a point), then $ax$ represents a point that is $a$ times as far to the right of 0 as is $x$.

We now consider the two points $ax$ and $bx$, where $a$ and $b$ are any two numbers, which may or may not be the same. The point $ax$ lies $a$ times as far to the right of 0 as $x$ does, and the point $bx$ lies $b$ times as far to the right of 0 as $x$ does. What do we mean by the point $ax + bx$? From our discussion of the arithmetic of points on a line in the first chapter, we know that we can reach the point $ax + bx$ by first moving along the line to point $ax$ and then moving along the line a distance $bx$ from this point to the point $ax + bx$. This series of steps is just equivalent to adding the distance $ax$ to the distance $bx$. To be specific, let $a$ be 4 and $b$ be 6. We reach the point $4x + 6x$ by moving a distance $4x$ to the right of 0 and then moving another distance $6x$ to the right of point $4x$. The total distance we have moved is thus $10x$ which is what we obtain when we add $4x$ to $6x$ so that $4x + 6x = (4 + 6)x = 10x$. This equation illustrates a basic algebraic rule: If every term in a sum has a common factor (in this example, $x$) we may remove this factor from each term. We then place it as a factor of the sum of the other factors of the individual terms in the sum by using parentheses. Returning now to the sum $ax + bx$, we have $(ax + bx) = (a + b)x$.

Here $x$ is a common factor and $a$ and $b$ are the other factors and the parentheses tell us that we must multiply each term by $x$.

We may now apply this rule to the sum of any number of terms such as $ax + bx + cx + dx$ and equate this expression to $(a + b + c + d)x$. This manipulation of the terms illustrates the distributive law of addition in algebra. The numbers $a$, $b$, $c$, $d$ may be integers (positive or negative), fractions, irrational, imaginary, or complex numbers. As long as $x$ is associated with real points on a line, it must be a real number, but we may dissociate $x$ from real points on a line and allow it to be any kind of number we please. With this change we enter the domain of algebra in its most general form.

The rules of algebra are exactly the same as those of arithmetic, with letters replacing numbers and with no restrictions on the numbers that letters may represent. We now restate the rules of arithmetic, as they apply to letters, without any reference to the meanings of these letters. If $x$ and $y$ represent two different sets of quantities (numbers, for example), then the sum of $x$ and $y$ is written as $x + y$ or $y + x$, which means that the order in which we add two algebraic quantities does not alter the sum. We extend this idea to any number of sets; we may use the three sets

$$x + y + z = x + z + y = x + (y + z) = (x + y) + z$$
$$= y + (x + z)$$

We use the parentheses to show that the sum is not altered if we group any two of the three sets and add these two together and then add the third one to the sum of the other two sets. We may also have subtractions within the sum. Thus,

$$(x + y - z) = (x + y) - z = (x - z) + y = -(-x - y + z)$$
$$= (x + y) - z$$

Note that we may place a minus sign in front of the parenthesis if we then change the sign of every term in the parentheses. This procedure is equivalent to multiplying the entire sum by $-1$. We can always check our algebra by replacing each of the letters by any number we please. If we replace $x$, $y$, $z$ by the numbers 2, 3, 6, respectively, we obtain the sum $(x + y - z) = 2 + 3 - 6 = -1$ and this is the value we get if we replace $x$, $y$, $z$ by 2, 3, 6 in all of the other expressions that we have equated to $(x + y - z)$. Thus, $-(-x - y + z) = -(-2 - 3 + 6) = 1$.

This is a very useful general rule which shows us very quickly whether our algebra is correct or not whenever we change one algebraic expression into another by any kind of algebraic manipulation. We give examples of this useful method of checking our algebra as we develop the material in this chapter.

We now write down the rules of addition and multiplication as they apply specifically to algebra. The law of addition is simply an extension of arithmetic addition to letters. In general, $ax + bx = (a + b)x$ as already stated where the numbers such as 2, 3, $a$, $b$ that stand in front of $x$ are called coefficients of $x$. The rule of addition is thus the following: To obtain the value of a sum of terms, each consisting of an $x$ with a coefficient, add up all the coefficients and place that sum as a coefficient of $x$. The coefficients may be positive, negative, real, imaginary, or complex, so that the final sum may be real, imaginary, or complex. If some of the coefficients are complex, we must add real parts to real parts of the coefficients and imaginary parts to imaginary parts. Thus,

$$4x + (5 + 2i)x - 5ix + 10x = (4 + 5 + 10)x + (2i - 5i)x$$

$$= 19x - 3ix = (19 - 3i)x$$

In our daily activities we constantly deal with quantities of different things to which we often assign letters to keep track of them just as we do in algebra, but we know that we cannot add the coefficients of these quantities to obtain anything meaningful. If we go shopping we carry a list such as $6a$, $4t$, $3b$, where $a$, $t$, $b$ stand for apples, tomatoes, and bananas, respectively, but adding the coefficients 6, 4, 3 to obtain 13 does not tell us anything except that we are to purchase 13 pieces of fruit. This restriction is a restatement of the old adage that "you cannot add apples and oranges," which has its counterpart in literary declaration that we should not mix metaphors. We can, of course, obtain a meaningful sum of the quantities in the above shopping list by replacing $a$, $t$, and $b$ by the item price of apples, tomatoes, and bananas. We may thus replace $a$ by $p_a$, the price of an apple, $t$ by $p_t$, the price of a tomato, and $b$ by $p_b$, the price of a banana, where $p$ stands for price and the subscripts a, t, b identify the kind of item referred to by the price. In using these symbols, we introduce the costs of the apples, tomatoes, and bananas, which are quantities of money and may therefore be added to give the total cost. We have gone into a detailed discussion of the addition process in this simple example to emphasize that letters that represent different kinds of quantities cannot be added.

As long as we are dealing with the same kinds of quantities, we may write down the sum $x + y + z$ where $x$ represents one group of such quantities, $y$ another group (of the same kind of quantities), and $z$ a third group of the same quantities. If we are counting the number of 10-year-old children in three different schools, and $x$, $y$, and $z$ are the numbers in these schools respectively, then $x + y + z$ is a meaningful sum; it is the total number of 10-year-olds in the three schools. We emphasize again that if we are dealing with numbers in the abstract, without any reference to any entities to which they refer, and we use letters to represent such numbers, we may add these letters in the usual arithmetic way. We return to this point later when we use algebra to represent physical entities.

With the law of addition of algebraic quantities and the restrictions on it understood, we now present the law of multiplication of algebraic quantities, which is also carried over from arithmetic (numbers) to letters. In numerical arithmetic, multiplication is designated with a diagonal cross $\times$ or with parentheses as in the example $2 \times 4 = (2)(4)$ or $2(4)$ but in algebra (the arithmetic of letters) this $\times$ is rarely used and the parentheses are used mostly when sums of quantities are multiplied together. Multiplication of two quantities such as that of $x$ and $y$ is understood when the two quantities are placed next to each other as illustrated by $xy$; a dot is sometimes placed between the two letters. Thus, $x \cdot y = xy = x(y) = (x)(y)$. The cross is not generally used because it may be confused with the letter $x$ which is used to represent algebraic quantities.

No restriction is placed on the entities that may be multiplied together in algebra. If $x$ represents a quantity of goods and $y$ the price, then the product $xy$ is a meaningful quantity, namely the cost of the goods. In fact, we shall see that we obtain new kinds of quantities only by multiplying different kinds of quantities together (or by dividing one kind of quantity by another kind). Thus, in the simple example above, "cost" comes from the product of "price" and "quantity of goods"; yet "cost," "price," and "quantity of goods" are different from each other. Even if we multiply two similar quantities together, the product is a different kind of quantity. Thus, $xx = (x)(x) = x^2$ is different in kind from $x$.

Just as we designate the multiplication of two different entities $x$ and $y$ by placing them next to each other as $xy$, we designate in the same way the multiplication of any algebraic quantity by a number. The product $2x$ means that we have two $x$'s so that $2x = x + x$ and $5x = x + x + x + x + x$. We may multiply a sum of similar entities by numbers or by other entities, and designate the multiplication by parentheses. Thus, the product of 3 and the sum $2x + 3y$ is written as $3(2x + 3y) = 6x + 9y$. The rule here is that we may place the multiplier outside the parentheses or multiply each term by the multiplier separately.

When written as $3(2x + 3y)$ we say that 3 is a factor of $(6x + 9y)$, but we can make 6 a factor or 9 a factor or any number we please a factor by placing that number outside the parentheses and then changing the coefficients of $x$ and $y$ appropriately to keep the expression correct. Thus,

$$(6x + 9y) = 3(2x + 3y) = 6(x + \tfrac{9}{6}y)$$
$$= 6(x + \tfrac{3}{2}y) = 10(\tfrac{6}{10}x + \tfrac{9}{10}y)$$
$$= 10(\tfrac{3}{5}x + \tfrac{9}{10}y)$$

and so on.

The same rules illustrated above apply to the multiplication of a sum of algebraic quantities by another algebraic quantity. Thus, $x(x + y + z) = xx + xy + xz = x^2 + xy + xz$. The way we have written the first term in the second equality is important for it introduces the concept of an algebraic entity raised to a power, which, in the above example, is the power 2 or the square of $x$. This same kind of designation applies to any power of $x$. The product $xxxxx$ is $x^5$ which we may also write as $xx^4 = x^2x^3$. Since we may raise any algebraic quantity to any power, we have the general rule that $xx \cdots x$, where the number of $x$ factors is $n$, is to be written as $x^n$. We may also write this quantity as $xx^{n-1}$ or $x^2x^{n-2}$ or $x^qx^{n-q}$, but these exponential relationships hold in algebra regardless of whether $n$ and $q$ are integers or not; they may indeed be any kind of numbers—integers, fractions, irrational numbers, or complex numbers.

The basic rule for multiplying quantities illustrated by these examples is that the product of different powers of a given entity equals that entity raised to the sum of the given powers. If we have $x^n$, $x^m$, $x^p$, $x^r$, where $n$, $m$, $p$, and $r$ are any numbers, then $x^nx^mx^px^r = x^{(n+m+p+r)}$. We may, of course, reverse this operation and write $x^{s+t} = x^sx^t$ or, as another example, $x^{a+ib}$ which equals $x^ax^{ib}$, where $a + ib$ is any complex number. Now we consider the product $x^{n-m}$ where $n$ and $m$ are any numbers. We then have $x^{n-m} = x^nx^{-m} = x^n/x^m$, so that a negative exponent means the reciprocal of the number ($x$, in this example, raised to the positive power). But $x^n/x^m$ means that we are dividing $x^n$ by $x^m$ so that a given power of $x$ divided by another power of $x$ is equivalent to the subtraction of the exponents as in arithmetic. We note a few more relationships which we carry over from arithmetic. Thus, $(xy)^n = x^ny^n$ and $(x/y)^n = x^n/y^n = x^n(1/y^n) = x^ny^{-n}$ which illustrates the rule that the product of any number of factors raised to any power equals the product of the individual factors raised to that power; an additional example is $(xyzu)^n = x^ny^nz^nu^n$, where $n$ may be any number.

We now note a few more rules and some restrictions on algebraic operations. Since $-x$ is the same as $(-1)(x)$, we have $(-x)^n = (-1)^nx^n$ and this quantity is $-(x^n)$ if $n$ is odd and $x^n$ if $n$ is even because $(-1)^0 = 1$, $(-1)^2 = (-1)(-1) = 1$ and so on, whereas $(-1)^1 = -1$, $(-1)^3 = (-1)(-1)(-1) = -1$, and so on. This relationship is also true of negative odd and negative even exponents so that $(-x)^{-n} = (x^{-n})$ if $n$ is even and equals $-(x)^{-n}$ if $n$ is odd. An important restriction in working with exponents is that a sum of terms raised to any power $n$ does not equal the sum of the individual terms raised to that power. Thus, $(x + y + z)^n \neq x^n + y^n + z^n$, where $\neq$ indicates that the two terms are unequal. Note also that $(x + y + z)^{-n}$ does not equal $(1/x + 1/y + 1/z)^n$. Following these rules and restrictions we can carry out any algebraic process we please without committing errors. But if we are in doubt about anything we have done, we can always substitute actual arbitrary numbers for the letters we have been using and check numerically the results of our algebra.

Consider now the product of two sums of different quantities such as $(x + y)$ and $(a + b)$, where $x$, $y$, $a$, $b$ may be any four numbers. This product, written as $(x + y)(a + b)$, must be a sum of four terms, each a different product of two of the four quantities $x$, $y$, $a$, $b$. This expression simply states the rule that each term in the sum $(x + y)$ must multiply each term of the sum $(a + b)$. We thus have $(x + y)(a + b)$ $= (x + y)a + (x + y)b = xa + ya + xb + yb$. In algebra one describes the equality of $xa + ya + xy + yb$ and $(x + y)(a + b)$ as factoring the sum $xa + ya + xb + yb$; $(x + y)$ and $(a + b)$ are then called the factors of the sum $xa + ya + xb + yb$. In elementary algebra classes a good deal of time is devoted to the study of factoring because factoring often simplifies an algebraic expression which makes the algebra easier. But it also plays an important role in the solution of algebraic equations which we discuss later in this chapter. A very useful example of factoring is given by the factors of $x^2 - y^2$ which are $(x + y)$ and $(x - y)$. Taking the product of these two factors, we obtain $(x + y)(x - y) = xx - xy + yx - yy = (x^2 - y^2)$ since the two middle terms cancel each other. But we can also factor $(x - y)$ by recalling that $x = (x^{1/2})^2$

and $y = (y^{1/2})^2$ so that we have $(x - y) = [(x^{1/2})^2 - (y^{1/2})^2]$ $= (x^{1/2} - y^{1/2})(x^{1/2} + y^{1/2})$. Hence, we can write $(x^2 - y^2)$ as a product of as many factors as we please. Thus,

$$(x^2 - y^2) = (x - y)(x + y)$$
$$= (x + y)(x^{1/2} + y^{1/2})(x^{1/2} - y^{1/2})$$
$$= (x + y)(x^{1/2} + y^{1/2})(x^{1/4} + y^{1/4})(x^{1/4} - y^{1/4})$$

and so on. The importance of reducing an algebraic expression to its simplest form by canceling out common factors is illustrated in the expression $(x^2 - y^2)/(x - y)$ if we allow $x$ to equal $y$. We then get $0/0$ which is meaningless. But we obtain a meaningful expression by placing $(x^2 - y^2)$ $= (x - y)(x + y)$. The $(x - y)$ in the numerator cancels the $(x - y)$ in the denominator and the fraction is equal to $2y$ for $x = y$.

We can easily discover whether a given algebraic expression is a factor of another one by simply dividing the first expression into the second one. This quotient and the first expression are then two factors of the second expression. As an example, consider the two expressions $(x + y)$ and $(x^2 + 2xy + y^2)$. Is the first expression a factor of the second expression? We obtain the answer by dividing the first expression into the second one which leads to the rule for division in algebra. This rule is similar to the one for long division in arithmetic which we learned in grade school. In arithmetic long division, we break the dividend (the number to be divided) into a sum of terms, each of which is divided separately by the divisor. The quotient is then the sum of these separate quotients. Thus, we divide 12 into 156 by writing 156 as $120 + (156 - 120)$ and dividing 12 first into 120, which gives 10, and then dividing 12 into the remainder $(156 - 120)$ or 36. This operation gives us 3 so that 156 divided by 12 is $10 + 3 = 13$. We can also do a different kind of long division in arithmetic by breaking up the divisor into a sum and then dividing this entire sum into the dividend. To divide 12 into 156 we write 12 as $10 + 2$ and note that it (the sum) can go 10 times into 156 leaving over $156 - (100 + 20) = 56 - 20 = 36$ into which $10 + 2$ goes exactly 3 times so that the total result of dividing $10 + 2$ or 12 into 156 is exactly 13. We must remember that though we may break up a divisor into a sum in division, we may not divide each term of this sum separately into the dividend and take the sum of these separate divisions as the quotient. In the above example, 156 divided by $10 + 2$ is not $156/10 + 156/2$ which is certainly not 13.

By applying these rules we now can divide one algebraic expression by another just as we divide one number by another in arithmetic long division. As an example we divide the expression $(x^3 + 3x^2 + 3x + 1)$ by $(x + 1)$, which we may write in various forms such as $(x^3 + 3x^2 + 3x + 1) \div (x + 1)$, $(x + 1)/(x^3 + 3x^2 + 3x + 1)^{1/2}$, $(x^3 + 3x^2 + 3x + 1)/(x + 1)$, $(x^3 + 3x^2 + 3x + 1)(x + 1)^{-1}$. Note that here both the dividend and the divisor are sums but our basic rules still apply. Considering the first term of the divisor and the dividend (the terms $x$ and $x^3$), we see that $x^3$ divided by $x$ is $x^2$ (exponents are subtracted when we divide). We therefore take $x^2$ as the first term of the quotient. We now multiply the divisor $(x + 1)$ by $x^2$ to obtain $x^3 + x^2$. This procedure shows us that $x^2$ is not the complete quotient for if it were, multiplying it by the divisor $(x + 1)$ would give us the complete dividend. To find out how much of the dividend is left over after this first step in our division operation, we subtract $x^3 + x^2$ from the dividend $x^3 + 3x^2 + 3x + 1$ to obtain

$$(x^3 + 3x^2 + 3x + 1) - (x^3 + x^2) = x^3 + 3x^2 + 3x + 1 - x^3 - x^2$$

$$= 2x^2 + 3x + 1$$

We now divide this remainder by $x + 1$, noting that $x$ goes into $2x^2$ just $2x$ times. We thus find that $2x$ is the second term in our quotient which we must add to $x^2$ to obtain the sum $x^2 + 2x$. But this is not the complete quotient either since we do not obtain the complete remainder $2x^2 + 3x + 1$ when we multiply $(x + 1)$ by $2x$. In fact, we obtain $(x + 1)(2x)$ $= 2x^2 + 2x$, which still leaves something over. We find this something by subtracting this product from our first remainder $2x^2 + 3x + 1$, and obtain $(2x^2 + 3x + 1) - (2x^2 + 2x)$ $= 2x^2 + 3x + 1 - 2x^2 - 2x = x + 1$. We complete our division now by dividing this final remainder by the divisor $(x + 1)$, which gives us 1. This quantity must be added to $x^2 + 2x$ to give the final quotient as $x^2 + 2x + 1$, with nothing left over.

We show each of the steps in this example of algebraic division by writing $x^3 + 3x^2 + 3x + 1$ as the sum $(x + 1)(x^2)$ $+ (x + 1)(2x) + (x + 1)(1)$. We see that this expression is a sum of terms, each of which contains the factor $(x + 1)$ so that

$$(x^3 + 3x^2 + 3x + 1) = x^2(x + 1) + 2x(x + 1) + (x + 1)$$
$$= (x + 1)(x^2 + 2x + 1)$$

We have thus factored $x^3 + 3x^2 + 3x + 1$ into the two factors $(x + 1)$ and $(x^2 + 2x + 1)$. Placing $x = 1$ in the first expression we obtain $(1 + 3 + 3 + 1) = 8$ and by doing this in $(x + 1)$ $(x^2 + 2x + 1)$ we obtain $(1 + )(1 + 2 + 1) = (2)(4) = 8$, which shows that the algebra is correct.

We now summarize the algebraic division process in a few steps: Divide the first term of the divisor into the first term of the dividend. Multiply this quotient by the divisor (by each term of the divisor) and subtract this product from the dividend. Now do exactly the same thing with this remainder and so on with each remainder until no remainder is left over or you arbitrarily stop with some remainder left over. Using these rules one may divide any algebraic expression by any other one, but, in general, some remainder will be left over. As an important and interesting example of algebraic division with an unending remainder, we consider the division of 1 by $(1 + x)$, which may be written as $1/(1 + x)$. Using our basic rule we divide 1 (the first term of the divisor) into 1 and obtain 1 as the first term of the quotient. We now multiply $(1 + x)$ by 1, obtain $1 + x$, and subtract this amount from 1, obtaining $1 - (1 + x) = 1 - 1 - x = -x$, which is the remainder. We now continue the division into the remainder by noting that the first term of $1 + x$ goes into $-x$ exactly $-x$ times so that $-x$ is the second term of the quotient which must be added to 1 to give $1 - x$ as the quotient up to this point in our division. We obtain the remainder after this step by multiplying $1 + x$ by $-x$ and subtracting this quantity from $-x$ as follows: $-x - (-x)(1 + x) = -x + x(1 + x)$ $= -x + x + x^2 = x^2$. The next term in our quotient is $x^2$ since the first term 1 of our dividend $1 + x$ goes into $x^2$ exactly $x^2$ times. Up to this point our quotient of $1/(1 + x)$ is $1 - x$ $+ x^2 \ldots$, where the dots stand for the additional terms of this unending (infinite) series if we continue step by step exactly as in our first three steps.

But we do not have to continue in this step-by-step procedure to find the terms that follow the third term $x^2$ in this series. The coefficients are either $+1$ or $-1$, with the coefficient $+1$ if the power of $x$ is even and $-1$ if the power is odd. We then have the remarkable result $1/(1 + x)$ $= (1 + x)^{-1} = 1 - x + x^2 - x^3 + x^4 - x^5 + \cdots$. This example shows the power and beauty of algebra, for the simple application of algebraic division permits us to write algebraic expressions as infinite sums of terms. But we obtain more than the infinite series for the fraction $1/(1 + x)$ by this process for, having obtained this infinite series, we obtain the infinite series for $1/(1 - x)$ by replacing $x$ in every term of the above series by $-x$. Thus,

$$1/(1+x) = 1/[1-(-x)]$$
$$= 1-(-x)+(-x)^2-(-x)^3+\cdots$$
$$= 1+x+x^2+x^3+\cdots$$

Indeed, we can obtain the series for

$$1/(1+x^n) = (1+x^n)^{-1} = 1-(x)^n+(x^n)^2$$
$$-(x^n)^3+(x^n)^4-\cdots$$
$$= 1-x^n+x^{2n}-x^{3n}+x^{4n}\cdots$$

We now note some other applications of the rules of arithmetic to algebra. We saw that if in a sum of numerical fractions the denominators of all the fractions are the same, we need only add the numerators, keeping their signs (positive or negative) unchanged, and then place this sum over the common denominator to obtain the sum of the fractions as in the example $1/5 - 2/5 + 4/5 = 3/5$. If, however, the denominators of the individual fractions are different, we first change them all to a common denominator and then add the numerators and place this sum over the common denominator. As a numerical example we consider the sum $1/2 - 2/3 + 4/5$. The smallest common denominator is 30 and we change each of the denominators into 30 by multiplying each fraction by 1 written as $15/15$ for the multiplication of $1/2$, $10/10$ for $2/3$, and $6/6$ for $4/5$. The sum then becomes $15/30 - 20/30 + 24/30 = (15 - 20 + 24)/30 = 19/30$.

Note that in multiplying a fraction by another fraction such as $15/15$ whose numerator and denominator are equal so that we are really multiplying by 1, we do not change the numerical value of the multiplied fraction—only its appearance. Note also the important rule, previously stated, that in multiplying any number of numerical fractions together, we multiply all the individual numerators together and all the individual denominators together. These same rules apply to the addition and multiplication of algebraic fractions as illustrated by the following examples.

Consider first the simple example $a/x + b/x - c/x$ where $a$, $b$, and $c$ may have any numerical values we please. Since $x$ is the common denominator here, we obtain the sum of these fractions by summing the numerators and placing this sum over $x$ so that $a/x + b/x - c/x = (a + b - c)/x$. If the denominators are different as in the example $a/x + b/y + c/z$, we obtain the common denominator $xyz$ by multiplying the first fraction by $yz/yz$, the second by $xz/xz$, and the third by $yx/yx$ so that

$$a/x + b/y + c/z = (a/x)(yz/yz) + (b/y)(xz/xz)$$
$$+ (c/z)(yx/yx)$$
$$= ayz/xyz + bxz/yxz + cyx/zyx$$
$$= (ayz + bxz + cyx)/xyz.$$

In another simple example we consider the sum

$$x/y - y/x = (x/y)(x/x) - (y/x)(y/y) = x^2/xy - y^2/xy$$
$$= (x^2 - y^2)/xy = (x - y)(x + y)/xy$$
$$= x(x + y)/xy - y(x + y)/xy$$
$$= (x + y)/y - (x + y)/x$$
$$= (x + y)(1/y - 1/x) = x/y - y/x.$$

We cite this example to illustrate how simple algebraic expressions can be altered in appearance without altering their meanings by manipulating them algebraically. Such manipulations can be very helpful in simplifying algebraic expressions and solving algebraic problems.

Just as we may change the sign of a number in arithmetic by multiplying it by $-1$, we may do the same thing to an algebraic fraction, keeping in mind that the minus sign may be applied to the entire fraction, or to the numerator, or to the denominator. Thus,

$$(-1)[(x - y)/(x + y)] = -(x - y)/(x + y)$$
$$= (-x + y)/(x + y)$$
$$= (x - y)/-(x + y)$$
$$= (x - y)/-x - y$$

All of these are equivalent ways of taking the negative of a fraction. Note also that if we raise a fraction to a given power, we must raise the numerator and the denominator to that power. Raising a fraction to a negative power means taking the reciprocal (inverting it) to the same power as in the example $[(x + y)/(x - y)]^{-n} = [(x - y)/(x + y)]^n$. The value of a fraction is altered if we add a given quantity to the numerator or to the denominator or if we add the same quantity to both the numerator and the denominator. Thus, $x/y \neq (x + a)/y \neq x/(y + a) \neq (x + a)/(y + a)$. If we do add a quantity to the numerator, we can always find a different quantity which, when added to the denominator, keeps the fraction the same. If we add $a$ to the numerator, what quantity must we add to the denominator to keep the fraction unaltered? Let this quantity be $z$ and see if we can discover how $z$ must be related to $a$ to keep the fraction equal to its original value. This means that if $x/y$ is the original fraction which we want to remain the same when we add $a$ to the numerator and $z$ to the denominator, we must have $x/y = (x + a)/(y + z)$, which is an algebraic equation for $z$ in its dependence on $x$, $y$, and $a$. If we solve this equation for $z$, we then have the answer we are seeking.

Before we show how to solve this equation for $z$, we discuss the general properties of algebraic equations which are probably the most useful intellectual and analytical devices ever created. Without the aid of such equations, science and technology would still be in their infancy and even such nonscientific adventures as political science, economics, sociology, etc., would be greatly hampered. Advanced mathematics itself has its roots in algebraic equations although it is often difficult to trace recent developments in mathematics back to their algebraic origins. Algebraic equations in science, particularly astronomy, physics, and chemistry, are important because the basic principles or laws in these disciplines are stated as algebraic equations or can be reduced to such equations. The task of the scientist is to understand the natural events he is studying, and he does this by applying to these events the appropriate natural laws. This application leads him to an equation or a set of equations which correlate to each other the quantities he wants to know. The solutions to these equations then give him the answers to the questions that these events present. Although the equations to which this analytical procedure leads may be quite complicated, they often can be reduced to much simpler algebraic equations which can be solved arithmetically. Indeed, all the modern computer can do is solve algebraic equations which it does arithmetically because the computer knows only arithmetic.

In dealing with algebraic equations we differentiate between equations and identities as illustrated by such relationships as $(x^2 - y^2) \equiv (x + y)(x - y)$ where the symbol "$\equiv$" means "identity." An identity is a correct statement no matter what numbers are substituted for the letters (which, in the above example, are $x$ and $y$) that are involved in the identity or in which one side of the identity can be changed exactly into the other side by algebraic manipulations. Thus, if we multiply the quantities in the two parentheses together on the right-hand side of the above identity, we obtain $x^2 - y^2$ so that we have $0 = 0$. In a true equation, one of its sides cannot be changed to the other no matter how we manipulate the two sides.

In working with equations of any kind, algebraic or otherwise, the basic rule that we must follow is never to do anything to either side of the equation that invalidates it. This means that we must treat each side equally in any manipulations we perform. Thus, we may add or subtract anything we please to or from one side of the equation if we do the same thing to the other side. This rule applies to all other arithmetic or algebraic operations we may perform on either side of an equation. These operations are multiplying (or dividing), raising to powers, taking roots, and taking logarithms. Other nonalgebraic operations can be applied to both sides of an equation without invalidating it.

With these rules understood we now go back to the equation $x/y = (x + a)/(y + z)$ that started this train of thought. We recall that $z$ is the quantity we must add to the denominator to keep the fraction $x/y$ unchanged if $a$ is added to the numerator $x$. Since we seek $z$, we must manipulate the above equation so that $z$ stands on one side of the equation and everything else is on the other side. We begin by inverting both sides of the equation which is equivalent to raising both sides to the power $-1$. Thus, we obtain

$$y/x = (y + z)/(x + a)$$

We now multiply both sides by $(x + a)$ so that we have

$$[(x + a)(y)]/x = [(x + a)(y + z)]/(x + a) = y + z$$

The $(x + a)$ in the numerator and the denominator cancel each other out. We now subtract $y$ from both sides of this last equation, obtaining $\{[(x + a)(y)]/x\} - y = z$. We can simplify the left-hand side by writing $y$ as $yx/x$ and combining it into a single fraction with the first term since both have the same denominator $x$. Thus, $[(x + a)(y) - (y)(x)]/x = z$, or

$$z = [(x)(y) + (a)(y) - (y)(x)]/x = ay/x$$

This expression tells us that if we add $a$ to $x$ in the numerator of the fraction $x/y$, we must add $ay/x$ to the denominator $y$ to keep the value of the fraction the same. To illustrate this with a numerical fraction we consider $3/5$ and add 4 to 3 so that the numerator becomes 7. We must now, using the value for $z$, add $(4 \times 5)/3$ to 5 in the denominator so that the fraction becomes $(7 \times 3)/(15 + 20) = 21/35 = 3/5$, which is our original fraction. We have treated this problem in considerable detail to illustrate how we may manipulate an algebraic equation and how we solve it for what is called the unknown, in this example, $z$. Here the given quantities are $x$, $y$, and $a$.

Every algebraic equation contains a set of quantities, which are the given quantities, and may be chosen arbitrarily, and one other quantity, the unknown, whose value can be found only by solving the equation. If two or more unknown quantities are present in an algebraic equation, two or more algebraic equations containing these unknowns must be set up to solve the problem or, alternatively stated, to determine the values of the unknowns as solutions of the equations.

Before we consider algebraic equations in more detail, we introduce the concept of the algebraic polynomial in a single variable or unknown which we designate as $x$ and which may represent a number, a set of numbers, or any other kind of entity. For the moment, however, we associate $x$ with numbers only, but, beyond that, we leave it undetermined.

We may raise $x$ to any power we please and multiply that power of $x$ by any number we please to obtain quantities such as $ax^0$, $bx^1$, $cx^2$, etc., where $a$, $b$, $c$, etc. are called coefficients, as previously defined.

We now consider sums of such multiples of different powers of $x$ such as $ax^0 + bx^1 + cx^2 + dx^3 + \cdots + ex^n$, where the dots stand for all the terms with successive powers of $x$ between 3 and $n$, where $n$ may be as large a positive integer as we please. Since $x^0 = 1$, and $x^1 = x$, the sum may be written as $a + bx + cx^2 + dx^3 + \cdots + ex^n$; it is called a polynomial. The prefix "poly" stands for many, that is, at least 2. Some of the coefficients in a polynomial may be zero so that the power associated with that coefficient will be missing. We also assign a "degree" to a polynomial which is given by $n$, the largest exponent or power that appears in the sum. If only the power 0 occurs, we have a zero-degree monomial (a single term) which is just a number (a constant) and does not depend on $x$ at all. It remains the same no matter what value we assign to $x$. If the polynomial contains the two terms $a + bx$, it is a first-degree or linear polynomial, which we can also write as $b[(a/b) + x]$.

The linear binomial does depend on $x$ in a very simple way. If $x$ is 1, the value of this binomial is $a + b$, and if $x$ is 2, its value is $a + 2b$ and so on. We may, of course, also assign negative values to $x$ so that for $x = -2$, the value of the binomial is $a - 2b$. We note that if $a$ and $b$ are positive, this first-degree binomial increases steadily as we introduce increasing values of $x$; it never gets smaller as long as we go from smaller to larger positive values of $x$. However, if we use negative values of $x$, the value of the polynomial does decrease, ultimately passing through 0 and becoming negative if $x$ becomes smaller (by smaller we mean that it is more negative so that, in this sense, $x = -5$ is smaller than $x = -4$)

than a certain negative value. The value of $x$ for which a polynomial (a binomial in this example) is zero is called a "root" of the polynomial; the roots of polynomials play very important roles in the theory of algebraic equations.

To obtain the root of the linear (first-degree) binomial in our example, we equate it to zero, thus obtaining the linear algebraic equation $a + bx = 0$. On subtracting $a$ from each side of the equation, we obtain $a + bx - a = -a$ or $bx = -a$. By dividing both sides of the equation by $b$, we have $bx/b = -a/b$. This expression gives us the quantity $x = -a/b$ as the root of the linear binomial $a + bx$. We also call $x = -a/b$ the root of the first-degree equation $bx + a = 0$ and note that this equation has only one root. No value of $x$ other than $-a/b$ can make the left-hand side of the equation vanish. As a numerical example of the linear binomial we consider it with coefficients $a = 4$ and $b = 7$; namely $4 + 7x$. Its root and that of its linear equation $4 + 7x = 0$ is $x = -4/7$. We note that we may also write this binomial as $7(4/7 + x)$ which leads to the equation $7(4/7 + x) = 0$. When we have a nonzero value multiplying the entire left-hand side of an equation we may divide through by it (the left- and right-hand sides) to obtain $4/7 + x = 0$, since 0 divided by 7 $(0/7)$ is still 0.

We now consider the second-degree or quadratic trinomial $a + bx + cx^2$ which we may also write in the form $c[a/c + (b/c)(x) + x^2]$. This is a very important polynomial because it appears in the mathematics of conic sections (the mathematics of the curves we get when we slice through a cone) and the mathematics of the orbits of planets. The equation associated with the quadratic trinomial is $a + bx + cx^2 = 0$ and finding its roots is called solving the quadratic equation for $x$.

Before we study the roots of the quadratic equation (analyze its solution), we consider in greater detail how the quadratic trinomial $a + bx + cx^2$ differs from the linear binomial $a + bx$. The striking difference, of course, is the presence of the additional term $cx^2$ in the quadratic equation; it accounts for differences between the numerical values of the linear binomial and those of the quadratic trinomial for various numerical values of $x$. We saw that the linear binomial equals 0 when $x = -a/b$ and that it increases steadily for all larger values of $x$ and decreases (becomes ever more negative) for all values of $x$ smaller than $-a/b$. Owing to this behavior of the linear binomial, we say that it is a monotonic function of $x$. Here we introduce the function concept for the first time, which we discuss in greater detail later.

Turning now to the quadratic trinomial, we see at once that the quadratic term $cx^2$ is positive or negative depending on whether $c$ is positive or negative (regardless of whether $x$ is positive or negative because $x^2$ is always positive even if $x$ is negative, because the square of a negative number is positive). This means that the term $cx^2$ in the quadratic prevents it from constantly increasing with increasing or decreasing $x$. Owing to this fact, the quadratic trinomial is not a monotonic function of $x$.

We can best illustrate how the quadratic trinomial behaves as a function of $x$ by taking a specific example such as $2 - 6x + 3x^2$, where we have placed $a = 2$, $b = -6$, and $c = 3$. For every value of $x$, this quadratic trinomial has a definite value such as 2 for $x = 0$, $-1$ for $x = 1$, 5 for $x = -1$. To avoid repeating the words "quadratic trinomial," we represent it by the letter $y$, writing $y = 2 - 6x + 3x^2$, and call $y$ an algebraic function of $x$. This statement simply means that for every numerical value of $x$, we obtain a single numerical value of $y$, noting that, depending on the choice of $x$, $y$ may be rational, irrational, real, or complex. An important question in algebra about the function $y$ is: Which value or values of $x$ make $y$ zero? As we have already noted, these are the roots of the quadratic trinomial; there are two such roots and we shall see later how we can easily obtain both roots from a general formula. But before we reveal this magic formula, we note that we can obtain the roots by trial and error. Since $y = 2$ for $x = 0$ and $y = -1$ for $x = 1$, it must be 0 for some value of $x$ between 0 and 1 and so we might try $x = 1/2 = 0.5$. This gives the value $y = 2 - 3 + 3/4 = -1 + 0.75 = -0.25$ and therefore less than 0. By increasing $x$ in very small steps, we can get as close to 0 as we please by this trial-and-error method. The exact correct value of $x$ that makes $y = 0$ is $1 - \sqrt{3}/3$. We can check that this is so by placing $x = 1 - \sqrt{3}/3$ in the trinomial $2 - 6x + 3x^2$ itself. We thus obtain

$$y = 2 - 6(1 - \sqrt{3}/3) + 3(1 - \sqrt{3}/3)^2$$

$$= 2 - 6 + (6/3)\sqrt{3} + 3(1 - 2\sqrt{3}/3 + 3/9)$$

$$= 2 - 6 + (6\sqrt{3})/3 + 3 - 2\sqrt{3} + 1$$

$$= 6 - 6 + 2\sqrt{3} - 2\sqrt{3} = 0$$

Thus, one root of the quadratic equation $3x^2 - 6x + 2 = 0$ is $1 - 1/\sqrt{3}$. We know, by examining the trinomial $y = 2 - 6x + 3x^2$ carefully, that a second value of $x$ makes $y = 0$ or, alternatively stated, that $y$ has another root because $y$ becomes negative for values of $x$ smaller than $1 - \sqrt{3}/3$ $= 1 - 1/\sqrt{3}$, but it does not continue decreasing (becoming more and more negative) as $x$ is decreased. In fact, we see that for $x = 0$, $y = 2$ so that it changes from a decreasing function of $x$ to an increasing one. It starts decreasing from 0 at $x = 1 - 1/\sqrt{3}$ to $-1$ at $x = +1$ and then increases steadily so that it must become 0 for some value of $x$ between $+1$ and $+2$, for which $y$ is $+2$. The second value of $x$ for which $y$ is 0 is $1 + 1/\sqrt{3}$ so the two roots of $3x^2 - 6x + 2$ are $1 - 1/\sqrt{3}$

and $1 + 1/\sqrt{3}$; we therefore write the two roots in a single expression as $x = 1 \pm 1/\sqrt{3}$ where $\pm$ means plus or minus. We see that the function $y$ is symmetrical about the value $x = 1$, increasing steadily for values of $x$ less than 1 and for values of $x$ larger than 1.

We can write the quadratic equation $3x^2 - 6x + 2$ in the form of a product of two factors to show its roots explicitly. Thus, $y = 3x^2 - 6x + 2 = 3[x - (1 - 1/\sqrt{3})][x - (1 + 1/\sqrt{3})]$ which shows us that if either bracket is 0, $y$ is 0, and this occurs for $x = 1 \pm 1/\sqrt{3}$. If we multiply out the two factors of $y$ we obtain $3x^2 - 6x + 2$ which shows that the two roots are $1 \pm 1/\sqrt{3}$.

We have analyzed the quadratic $3x^2 - 6x + 2$ in considerable detail because if we know all the properties of any specific quadratic equation we can understand any quadratic equation regardless of the numerical values of the coefficients $a$, $b$, $c$. To illustrate this point, we consider the general quadratic equation $y = ax^2 + bx + c$, and see if we can factor it to show its two roots. Since we do not know these two roots at this point we call them $x_1$ and $x_2$. The values of these variables, of course, depend on the values of $a$, $b$, and $c$. We now write $y = 4a[x^2 + (b/a)(x) + c/a] = a(x - x_1)(x - x_2)$, showing that $y = 0$ for $x = x_1$ and $x = x_2$. By multiplying out the two factors on the right, we obtain $a[x^2 + (b/a)x + c/a] = a[x^2 - (x_1 + x_2)x + x_1x_2]$.

When we have an equation of this sort, it is really an identity and must therefore be true for all values of $x$. No matter what value we insert for $x$ in the equation, the left-hand side of the identity must still equal the right-hand side. We first place $x = 0$ on the left and on the right so that the terms containing $x$ vanish. We then have left over the identity $ac/a = ax_1x_2$; this tells us that the product of the two roots is $c/a$. If we now replace $x_1x_2$ on the right side of the above equation by $c/a$ we have

$$a(x^2 + xb/a + c/a) = a[x^2 - (x_1 + x_2)x + c/a]$$

or

$$a[x^2 + (b/a)(x)] = a[x^2 - (x_1 + x_2)x]$$

since the last terms on both sides of the equation are equal and therefore cancel each other. We therefore have $a[x^2 + (b/a)(x)] = a[x^2 - (x_1 + x_2)x]$. Since $x$ is present in each term on both sides, we may cancel $x$ out of each term to obtain $a(x + b/a) = ax - a(x_1 + x_2)$. Placing $x = 0$ on both sides of this identity we obtain $b = -a(x_1 + x_2)$ or $(x_1 + x_2) = -b/a$. This gives us the sum of the two roots so that we now have their product and their sum and, hence, enough information to determine each root completely in terms of the coefficients $a$, $b$, $c$.

This is a fairly straightforward algebraic exercise. We have the two equations $x_1 x_2 = c/a$ and $(x_1 + x_2) = -b/a$ for the two unknowns $x_1$ and $x_2$ so that we may find their values precisely. The general rule for finding unknowns is that we must have as many equations as we have unknowns so that by having the two equations above for $x_1$ and $x_2$, we can find $x_1$ and $x_2$ separately. Two such equations for two unknowns are called simultaneous equations which we discuss in greater detail later.

Returning now to our two simultaneous equations $x_1 x_2 = c/a$ and $(x_1 + x_2) = -b/a$ for $x_1$ and $x_2$ we note first that we may write $x_1 = -b/2a + q$ and $x_2 = -b/2a - q$ which gives us the correct value $x_1 + x_2$, namely $-b/a$, and where $q$ is a quantity to be determined. Since $x_1$ (or $x_2$) is a root of the quadratic $ax^2 + bx + c$, we must obtain 0 when we replace $x$ in the quadratic by $x_1$, which is $-b/2a + q$ (or by $x_2$). After doing this we obtain $a(-b/2a + q)^2 + b(-b/2a + q) + c$

which must equal 0 so that we thus obtain an equation for $q$ from which $q$ can be found. We have, on carrying out the algebraic operations in this sum,

$$a\left[\frac{b^2}{4a^2} - \frac{b}{a}(q) + q^2\right] - \frac{b^2}{2a} + bq + c = 0$$

or

$$\frac{b^2}{4a} - bq + aq^2 - \frac{b^2}{2a} + bq + c = 0$$

or, on combining common terms

$$aq^2 - \frac{b^2}{4a} + c = 0$$

Thus,

$$aq^2 = \frac{b^2}{4a} - c \quad \text{or} \quad q^2 = \frac{b^2}{4a^2} - \frac{c}{a} = \frac{1}{4a^2}(b^2 - 4ac)$$

so that $q = \pm(1/2a)\sqrt{b^2 - 4ac}$. Since we may place $q$ equal to either the positive or negative of the above quantity, we obtain the two roots

$$x_1 = -b/2a + (1/2a)\sqrt{b^2 - 4ac}$$

and

$$x_2 = -b/2a - (1/2a)\sqrt{b^2 - 4ac}$$

We may use a single expression to represent both roots as follows: $x_1$ (or $x_2$) $= (1/2a)(-b \pm \sqrt{b^2 - 4ac})$.

A more direct way of obtaining the two roots of $ax^2 + bx + c = 0$ is by an algebraic trick called completing the square. In this procedure we add and subtract the same quantity to the quadratic which permits us to factor it immediately. We first divide the quadratic equation through by $a$ to obtain the equation $x^2 + (b/a)(x) + c/a = 0$ and we then add and subtract the term $b^2/4a^2$ on the left-hand side. We thus obtain the same equation in the form

$$\left[ x^2 + \frac{b}{a}(x) + \frac{b^2}{4a^2} \right] + \frac{c}{a} - \frac{b^2}{4a^2} = 0$$

The quadratic expression in the brackets is the square of the quantity $(x + b/2a)$. We may therefore write the equation as $(x + b/2a)^2 - [\sqrt{b^2/4a^2 - c/a}]^2 = 0$ so that the left-hand side is the difference of two squares such as $p^2 - q^2$ which equals $(p + q)(p - q)$ as previously explained. We may therefore write the left-hand side of our equation as the product of the sum and difference of the two quantities that are squared. We thus obtain the equation

$$\left[ x + \frac{b}{2a} - \sqrt{\left( \frac{b^2}{4a^2} - \frac{c}{a} \right)} \right]\left[ x + \frac{b}{2a} + \sqrt{\left( \frac{b^2}{4a^2} - \frac{c}{a} \right)} \right] = 0$$

This equation has the two roots

$$x_1 = -b/2a + (1/2a)\sqrt{b^2 - 4ac}$$

and

$$x_2 = -b/2a - (1/2a)\sqrt{b^2 - 4ac}.$$

The quantity $b^2 - 4ac$ under the square root sign, called the discriminant of the equation, plays an important role in the algebra of quadratic equations. If the discriminant is positive $(b^2 > 4ac)$, the quadratic has two different real roots. If $b^2 = 4ac$ (the discriminant is 0), the two roots are real and equal. Finally, if $b^2 < 4ac$ (the discriminant is negative), the roots are complex and they are complex conjugates of each other. Considering this last case we write the discriminant in the form $-(4ac - b^2)$ so that we have

$$x_1, x_2 = -\frac{b}{2a} \pm \frac{1}{2a}\sqrt{-(4ac - b^2)}$$

$$= -\frac{b}{2a} \pm \frac{1}{2a}(\sqrt{-1})\sqrt{4ac - b^2}$$

$$= -\frac{b}{2a} \pm i\left(\frac{1}{2a}\right)\sqrt{4ac - b^2}$$

If we take $x_1 = -b/2a + (i/2a)\sqrt{4ac - b}$, then $x_2 = -b/2a - (i/2a)\sqrt{4ac - b}$ and we see that $x_2^* = x_1$, so that $x_2$ is the complex conjugate of $x_1$. A quadratic cannot have one real and one complex root because the sum of the roots is real (equal to $-b/a$) which would not be so if one root were real and one root complex. We note here a simple but important example of a quadratic equation $x^2 + 1 = 0$. This equation has the two roots $\pm\sqrt{-1} = \pm i$; this relationship leads to the study of complex numbers previously discussed.

Leaving the quadratic trinomial, we can go on to polynomials of higher degrees such as cubics (third degree), quartics (fourth degree), and so on. The general polynomial is of the $n$th degree, where $n$ may be any integer, but a general formula for the roots of a polynomial of a degree higher than 3 (the cubic) cannot be written. Although the formula for the three roots of a cubic is known as the Cardan formula, it is extremely complex and difficult to manipulate. Fortunately, most phenomena of interest in nature can be described by quadratic equations which can be solved by the quadratic formula as described previously.

Without discussing in any detail the polynomials of degree higher than 2, where by "the degree of the polynomial" we mean the highest power of the unknown $x$ that appears in the polynomial, we note some general properties. We write the polynomial of degree $n$ as the sum $a_n x^n + a_{n-1} x^{n-1} + \cdots + a_1 x + a_0$, where the dots stand for all the terms with powers of $x$ between $n - 1$ and 1 such as $a_5 x^5$, etc. The subscript such as $n - 1$ or 1 attached to each coefficient is merely an identifying number and is taken to be equal numerically to the exponent of $x$ which the coefficient multiplies. This way of identifying or labeling the coefficient of a term in a series of terms is standard in algebra and other branches of mathematics. Considering the $n$th-degree polynomial described above, we note that it must have $n$ roots, some of which may be equal (identical) to each other. This conclusion follows immediately from the observation that an $n$th-degree polynomial can be written as a product of exactly $n$ linear (first-degree) factors. Thus, we must have

$$a_n x^n + a_{n-1} x^{n-1} + \cdots + a_1 x^1 + a_0$$

$$= a_n (x - x_1)(x - x_2) \cdots (x - x_n)$$

where the dots on the right-hand side stand for all the factors in which the constant terms have all the subscripts from 3 to $n - 1$ such as $(x - x_3)$, $(x - x_4)$, etc. The equality of the polynomial and this kind of product of $n$ factors, which is zero if $x$ equals any one of the numbers $x_1, x_2, \ldots, x_n$, shows us that the $n$th-degree polynomial has exactly the $n$ roots $x_1, x_2, \ldots, x_n$. This statement is known as the fundamental theorem of algebra.

We can discover some of the properties of the roots $x_1, x_2, \ldots, x_n$ by multiplying out the $n$ factors $(x - x_1)(x - x_2) \cdots (x - x_n)$ of the polynomial and then equating the coefficient of each power of $x$ on the left-hand side of the above identity to the coefficient of the same power of $x$ on the right-hand side. We then find that the product of all the roots (the product $x_1 x_2 \cdots x_n$) equals $\pm a_0 / a_n$ depending on whether $n$ is even or odd, the sum of all the roots $x_1 + x_2 + \cdots + x_n = -a_{n-1}/a_n$, and so on. We do not pursue this idea any further here since this is a specialized branch of algebra which is beyond the scope of this book. However, we note one more property of the roots. We presented a special case of this property in our discussion of the quadratic trinomial. If one of the $n$ roots of the $n$th-degree polynomial is complex, the complex conjugate of this root must also be a root. The sum of all the roots must be the real number $-a_{n-1}/a_n$ (all the coefficients $a_n$, $a_{n-1}$, etc., are real) and that would not be true if complex roots did not come in pairs whose imaginary parts are the negative of each other (as in complex conjugates) so that these imaginary parts cancel each other when added.

Before leaving the general polynomial of the $n$th degree, we consider a very special case—namely $x^n - 1$—a polynomial in which all the coefficients between $a_n$ and $a_0$ are zero, and the two remaining coefficients $a_n$ and $a_0$ are $+1$ and $-1$. If we write this polynomial as an equation $x^n = 1$, we see that we obtain the solution by taking the $n$th root of both sides, that is, raising both sides to the $1/n$th power. Thus, $(x^n)^{1/n} = (1)^{1/n} = \sqrt[n]{1}$ or $x = \sqrt[n]{1}$ and we are faced with the problem of finding the $n$ roots of 1. For $n$ equal to 2 we have the two roots $+1$ and $-1$, both of which, when squared, give 1 as required. This is obvious by inspection but for values of $n$ larger than 2, we cannot obtain the roots in so simple a manner, and so we defer a detailed discussion of these roots for a later chapter, noting that the $n$ roots of 1 play an important role in the algebra of complex numbers and in trigonometry.

Though we defer the general discussion of the $n$ roots of 1, we pursue it one step further to deducing the three roots

of 1 which shows how all higher roots can be obtained by
simple algebraic manipulations. Our problem then is to solve
the equation $x^3 - 1 = 0$, which we do by factoring the left
side into the three factors $(x - x_1)$, $(x - x_2)$, $(x - x_3)$ where
$x_1$, $x_2$, $x_3$ are the three roots. We thus have $(x - x_1)(x - x_2)$
$(x - x_3) = 0$ and on multiplying out the left-hand side of the
expression, we obtain the following expression:

$$x^3 - (x_1 + x_2 + x_3)(x^2) + (x_1 x_2 + x_1 x_3 + x_2 x_3 (x) - x_1 x_2 x_3 = 0$$

Comparing the left-hand side of this form of our equation
with $x^3 - 1$ (the left-hand side of the equation we started
with), we must place the coefficients of $x^2$ and $x$ equal to
$0 - 1$ to make the two left-hand sides equal to each other as
they must be. We thus obtain the three equations

$$x_1 + x_2 + x_3 = 0, \ x_1 x_2 + x_1 x_3 + x_2 x_3 = 0, \text{ and } x_1 x_2 x_3 = 1$$

We know that one of the three cube roots of 1 must be 1
because $1^3 = 1$ (every cube root of 1 must give 1 when cubed)
and this is the only real cube root of 1 because it is the only
real number which, when cubed, gives 1.

Knowing that 1 is a cube root of 1, we may place one
of the three quantities $x_1, x_2, x_3$ equal to 1 and we choose
$x_1 = 1$. We therefore place $x_1 = 1$ in each of the three previous
equations for the roots to obtain the equations $1 + x_2 + x_3 = 0$,
$x_2 + x_3 + x_2 x_3 = 0$, and $x_2 x_3 = 1$. Since $x_2 x_3$ equals 1, the first
and second equations are identical. We are thus left with the
two equations $x_2 + x_3 + 1 = 0$ and $x_2 x_3 = 1$. Multiplying this
first equation by $x_2$ and again replacing $x_2 x_3$ by 1, we obtain
the equation $x_2^2 + x_2 + 1 = 0$. This is the quadratic equation
for the root $x_2$ whose two solutions as given by the basic

quadratic formula with $a = b = c = 1$ are

$$x_2 = -\tfrac{1}{2} \pm \tfrac{1}{2}\sqrt{1 - 4} = -\tfrac{1}{2} \pm \tfrac{1}{2}\sqrt{-3}$$

$$= -\tfrac{1}{2} \pm \tfrac{1}{2}\sqrt{-1}\sqrt{-3} = -\tfrac{1}{2} \pm \frac{i\sqrt{3}}{2}$$

We thus find that the three cube roots of 1 are 1, $(-1/2 + i\sqrt{3}/2)$, and $(-1/2 - i\sqrt{3}/2) = -1/2(1 + i\sqrt{3})$. If we cube $-1/2(1 - i\sqrt{3})$ or $-1/2(1 + i\sqrt{3})$, we obtain 1 so that these two complex conjugate quantities are indeed cube roots of 1.

We have discussed the cube roots of 1 in some detail to show the algebra of complex quantities and to illustrate the manipulation of algebraic equations. Later in our discussion of trigonometry we shall see that the various roots of 1 can be obtained by inspection, from the properties of a circle, without solving any equations. But obtaining the cube roots of 1 algebraically, as we did, demonstrates a very useful mathematical technique, namely that of reducing a complex problem to a simpler one by the appropriate application of algebra.

We complete our discussion of this phase of algebra, which encompasses what is generally called intermediate algebra, by showing how useful and powerful it is in solving certain elementary problems and we start with continued fractions, which we introduced in Chapter 2. We consider the continued fraction

$$\cfrac{1}{x + \cfrac{1}{x + \cfrac{1}{x + 1 \cdots}}}$$

and we use the letter $y$ to represent it. We thus have

$$y = \cfrac{1}{x + \cfrac{1}{x + \cfrac{1}{x + \cdots}}}$$

For every value of $x$ we obtain a definite value of $y$ so that $y$ is a continuous function of $x$. Can we find a simpler expression for $y$ (in its dependence on $x$) rather than the puzzling and mysterious form of a continued fraction? We can, indeed, by applying a few simple algebraic operations (manipulations) to this function of $x$. We first invert the left-hand side and the right-hand side of the equation to obtain

$$1/y = x + \cfrac{1}{x + \cfrac{1}{x + \cfrac{1}{x + \cfrac{1}{x + \cdots}}}}$$

which is permitted since in dealing with an equation we may do anything we please to one side of the equation, provided we do exactly the same thing to the other side. Here we have turned both sides upside down. Upon examining the term that follows $x$ on the right-hand side, we see that it is exactly the same as the right-hand side of the original equation and therefore equal to $y$. We thus have $1/y = x + y$, and, on multiplying both sides by $y$, we obtain $y/y = xy + y^2$, which leads to the quadratic equation $1 = xy + y^2$ for $y$ or $y^2 + xy - 1 = 0$. Noting that the coefficients $a$, $b$, $c$ in this quadratic are 1, $x$, $-1$, respectively, we have, from the quadratic formula for the two roots,

$$y = -\frac{x}{2} \pm \frac{1}{2}\sqrt{(x^2 + 4)} = \frac{1}{2}[-x \pm \sqrt{(x^2 + 4)}]$$

Thus, for $x = 1$ the continued fraction becomes

$$y = \cfrac{1}{1 + \cfrac{1}{1 + \cfrac{1}{1 + \cdots}}}$$

which we cannot evaluate as it stands but we obtain its value by substituting 1 for $x$ in the quadratic solution above. We thus obtain $y = 1/2[-1 + \sqrt{(1 + 4)}] = (1/2)(\sqrt{5} - 1)$, where we must use the $+$ sign in front of the square root, not the minus sign, because the value of the continued fraction must be positive.

As another example of this algebraic technique of obtaining quick answers to problems that seem unanswerable at first sight, we consider the following infinite sum $y = 1 + x + x^2 + x^3 + \cdots$. This is called an infinite power series since it goes on forever with ever-increasing powers of $x$. To find the value of this infinite geometric series, as it is called, we write

$$y = 1 + (x + x^2 + x^3 + \cdots) = 1 + x(1 + x + x^2 + \cdots)$$

where we have taken a factor $x$ out of each term in the parentheses following 1 on the right-hand side. We see at once that the terms now in the parentheses are exactly those in our original sum and, hence, equal to $y$. We thus obtain the equation $y = 1 + xy$, which leads to the equation $y - xy$

$= 1$ or $y(1 - x) = 1$. On dividing both sides by $(1 - x)$ we obtain $y(1 - x)/(1 - x) = 1/(1 - x)$ or $y = 1/(1 - x)$. This is called the sum of the geometric series. This formula gives us the correct value for the sum only if $x$ is smaller than 1. For values of $x$ larger than 1 the sum is infinite. For all negative values of $x$, however, the formula gives us the correct sum. The terms of the series, then, alternate between positive and negative values and the series always remains finite. We obtained this series in Chapter 2 by division.

## NOTES

[1] Morris Kline, *Mathematics for the Nonmathematician.* New York: Dover, 1985, p. 94.

[2] *Ibid.,* p. 94.

[3] *Ibid.,* p. 95.

[4] *Ibid.,* p. 96.

[5] W. W. Rouse Ball, *A Short Account of the History of Mathematics.* New York: Dover, 1960, p. 103.

[6] *Ibid.,* p. 103.

# CHAPTER 4

# Graphic Algebra

*It is impossible not to feel stirred at the thought of emotions of men at certain historic moments of adventure and discovery—Columbus when he first saw the Western shore, Pizarro when he stared at the Pacific Ocean, Franklin when the electric spark came from the string of his kite, Galileo when he first turned his telescope to the heavens. Such moments are also granted to students in the abstract regions of thought, and high among them must be placed the morning when Descartes lay in bed and invented the method of coordinate geometry.*
—ALFRED NORTH WHITEHEAD

Up to this point in our description and analysis of algebra and algebraic processes we have used only words and symbols to develop our ideas and to explain the mathematics of algebra but in this chapter we turn to the pictorial or graphic representations of algebra that mathematicians have introduced over the years. The concept of the graph is familiar to all of us because we see graphs in newspapers, magazines, newsletters, television, etc., whether the subject matter involved is industry, politics, business, sports, health, environment, or climate. These graphs are generally used to reveal to the reader at a glance the trend of events of one kind or another, without presenting any relationship of these events to anything else. Thus, the fluctuations of the prices of stocks over a span of time, as represented by the changes in the Dow Jones average, are plotted from day to day or month to month, and such graphs are regular features of the financial pages of every newspaper. Nothing in such plots, however, reveals anything in the nature of a causal relationship between the Dow Jones average and any external agent or set of events. Such graphs teach us very little beyond showing that such things as stock prices fluctuate from day to day.

An examination of the nature of a graph quickly reveals to us whether, in any particular instance, it is meaningful (carries substantive information) or is an empty pictorial exercise. The basic elements in a graph are two intersecting lines, one an east-west line, which we call the horizontal axis or abscissa, and the other a north-south line, called the vertical axis or the ordinate of the graph. The point of intersection of the ordinate and abscissa is called the origin of the graph and is labeled 0. One notes in any specific graph that there are two sets of numbers, one set along the abscissa and the other set along the ordinate. If these two sets of numbers are not related to each other in any causal way, the graph has very little meaning but if the two sets of numbers are related in some intimate way, the graph reveals on inspection some kind of basic truth or even a basic principle. In general, the numbers along the abscissa and those along the ordinate do not represent or refer to the same kind of entity. In the Dow Jones graph the numbers along the vertical (north-south) axis or the ordinate are prices (of stocks) and the numbers along the horizontal (east-west) axis or abscissa are dates or times. We thus have a graph that relates price to time. However, no causal relation exists between price and time. The Dow Jones average is not altered by time even though the average and the time are related to each other by the graph. The graph in this example is merely a data bank, presented in the form of a curve, every point of which gives us the Dow Jones average at some particular date, but this curve tells us nothing about why the Dow Jones average has that value at that particular date or time.

As against graphs of the sort represented by the time fluctuations of the Dow Jones average of stock prices, we consider graphs which do indicate a causal relationship between two different sets of quantities but, which, in general, do not reveal this relationship. Thus, a plot of the annual yield of wheat per acre against the annual rainfall in our midwestern wheat-growing states shows a definite correlation between the two. If the yield is plotted along the ordinate axis and the rainfall along the abscissa axis, the points of the wheat-yield rainfall plane (each representing a yield and rainfall) lie on a curve that increases from left to right. From this curve we conclude that the annual rainfall and wheat yield are causally related so that increased rainfall, up to a certain point, results in increased yields. But knowing this alone is not the same as knowing why increased rainfall results in increased yield. Mathematicians call this kind of correlation a functional relationship and say that the annual wheat yield of an acre is a function of (dependent on) the annual rainfall.

With this introduction of the concept of the function, we discuss it in its most general form and go beyond such specific examples as stock prices and wheat yields. We say that two sets of quantities are functionally related if, given a quantity in one set, a relationship exists between the two sets which leads us to a definite quantity in the second set. As a specific example we place all the even integers in one of our two sets (which we call the $y$ set) and all the odd integers in the other set (which we call the $x$ set) and note that a given integer in the $y$ set is obtained from a definite integer in the $x$ set by multiplying this $x$ integer by 2. We may then state this functional relationship algebraically as $y = 2x$, noting that $x$ and $y$ are limited to integers. We can plot this relationship by using the abscissa in our graph as our $x$ axis and the ordinate as our $y$ axis. The functional relationship $y = 2x$

gives us a discrete set of points on the $x$–$y$ plane [the plane determined by the horizontal (east–west) $x$ axis and the vertical (north–south) $y$ axis]. Each such discrete point in the plane is identified by the two integers $x$ and $2x = y$. Thus, the point $(3, 6)$ in the plane is 3 steps to the right of the origin and 6 steps above the origin. The integers 3 and 6 are called the coordinates of the point in the plane just described.

This arithmetic example of a functional relationship between two sets of quantities and the two examples of graphs (stock prices and wheat yields) previously given are too specialized to reveal certain general features of functional relationships which we now consider, but which cited examples may or may not possess. Returning to the graph of stock prices we see that the relationship between a stock price and a date is not one-to-one; the same price of a given stock occurs for many different dates because of the fluctuations of stock prices. Moreover, nothing in the graph or in any imaginable algebraic formula enables us to assign a price to any stock at any given date; the price and date are completely unrelated and any correspondence is purely accidental. In our second example of rainfall and wheat yield per acre, the plotted quantities are not precisely related to each other by a rigid algebraic formula but approximately interrelated by an empirical formula that stems from our agricultural experience. In the third example, the functional relationship is precise, given by an exact algebraic formula.

A few other important features of functional relationships are noteworthy, one of which we have already mentioned: the uniqueness property or the one-to-one correspondence of the two sets of quantities that are functionally related. We say that the function is unique if one and only one value of the function corresponds to a single value of the variable quantity. This is certainly not true of the first two examples because on any given future date the Dow Jones average may have a range of different values, no one of which can be deduced from the graph. Similarly, we can have a range of wheat-yield values for any given value of the rainfall since this yield depends on other conditions of a climatic nature and on the properties of the soil. The stock prices and the wheat yields per acre are functions of many different variables and so a unique relationship between either of these entities and a single variable does not exist.

In the algebraic functional relationship $y = 2x$, the correspondence between any given $x$ value and the resulting $y$ value is unique—it is one-to-one so that for each given $x$, there is only one $y$ and for each given $y$ only one $x$ exists. We shall see that in mathematics both single-valued (one-to-one) and multiple-valued (many-to-one) relationships exist.

Another important question concerning functional relationships that must be considered is whether the function is continuous or not. By continuous we mean that for every value of the variable a definite value of the function exists. From this definition of continuity of a function we see that our algebraic function $y = 2x$, as defined, is not continuous because it is defined only if $x$ is an odd integer. If we plot $y = 2x$ in our $x$-$y$ plane, we do not obtain a continuous curve but an infinite set of disconnected points. If we allow $x$ to take on any numerical value, $y$ is a continuous function of $x$ and its plot in the $x$-$y$ plane is an unbroken straight line. In this chapter on graphic algebra we consider only continuous functions.

With this preliminary discussion of the general properties of functions we proceed to a more detailed discussion of the graphics of algebraic functions of various degrees—linear, quadratic, cubic, etc., which we have already discussed to some extent. In considering a quantity $y$ as a function of (depending on) another quantity $x$, which we write symbolically as $y = f(x)$ [where the symbol $f(\ )$ means "the function of" whatever stands in the parentheses] we may treat $y$ and $x$ purely as numbers without attaching any physical quantity or significance to them. In other words, they are not assumed to be measurable. The functional relationship between $y$ and $x$ is then entirely arithmetic with no physical significance. We could very well study functions from such an arithmetic point and learn all about them that way but it is much more interesting to assign measurable properties to $y$ and $x$ so that we can picture them as things we can experience. The simplest such physical entity is distance which is an important element of our daily lives. The quantities $y$ and $x$ are to be distances with the distance $y$ a function of (depending on or varying with) the distance $x$.

We now picture ourselves as walking along an east–west road (along a circle of latitude) from some fixed point 0 which is at sea level. The east–west straight line which passes through 0 and each point of which is at sea level is our abscissa; our distance from 0 as measured along this line (our $x$ axis) is our $x$ distance or, alternatively, our $x$ coordinate. This coordinate $x$ is positive if we are to the right of 0 and negative if we are to the left of 0. Although our road is due east–west, it is hilly with elevations and dips, which are to be measured vertically along a vertical pole that passes exactly vertically through the point 0. This vertical pole is our $y$ axis or ordinate; it extends high above the Earth and deep into its surface. If we are at any point on the road which is exactly at sea level, our $y$ coordinate (our elevation) is 0. If we are on the side of a hill above sea level on our road, our $y$ coordinate (our elevation) is positive; if our road dips below sea level, our $y$ coordinate (our distance below sea level) is negative. We see then that as we walk along the road our $x$

and $y$ coordinates change, and the relationship between $y$ and $x$ depends on the terrain, which can be described by the functional relationship between $y$ and $x$ at any point on the road. This relationship will, of course, vary from point to point since our $y$ coordinate can change from point to point.

If we always walk eastward, our eastward motion will never change so that our $x$ coordinate will continually increase. Our $y$ coordinate, however, may vary from point to point, at times increasing, at times remaining the same, and at still other times decreasing; all of these possibilities are described by the kind of function $y$ is of $x$. That means that $y = f(x)$ is really a description of the terrain or the hills and valleys in the road. Although the eastward direction of our walk does not change, our up and down direction does change; that change is determined by how $y$ changes as $x$ increases.

We may now describe all of these ideas by introducing a single distance which contains both the $x$ and $y$ data about our walk and where we are on the road. We draw a line from the origin 0 to our position and picture this line following us as we move along the road. Not only does its length increase from point to point but its direction changes as we walk along so that the end of it that is figuratively attached to us wiggles up and down as we walk along the road. The length of this line gives our direct distance from 0, which depends on our $x$ and $y$ coordinates. The direction of this line also depends on these $x$ and $y$ values so that the length and direction of this line, which we may call $r$, is a function of both $x$ and $y$, as expressed in the equation $r = f(x, y)$. This is a quantity which is a function of two different variables $x$ and $y$. We return to this very important concept in a later chapter.

We now return to $y$ and $x$ and study various simple functional relationships and see how these can be understood by our simple graphical methods. The simplest example of $y = f(x)$ is $y = a$ where $a$ is a constant. Note that since $y$ is a distance, $a$ is also a distance, expressed in the same units of distance as $y$, so that if $y$ is expressed in centimeters (cm), $x$ is also expressed in the same units of measurement, in accordance with our rule that the entities on both sides of an equation must be the same kind of entities. Since $y$ is our vertical distance from sea level, the equation $y = a$ simply means that we are always $a$ units of distance above sea level as we walk eastward. The equation $y = 0$ means we are always at sea level and that the road has no hills or valleys. If $y$ equals some value different from 0, for example, $y = 3$, we know that the road is on an east–west plateau 3 units of distance above sea level. If we walk eastward on this road, our $x$ coordinate increases steadily while our $y$ coordinate remains constant. We represent all such possibilities for a road with no hills or dips by the equation $y = a$ where $a$ is any fixed number we please. The profile of this road plotted in our $x$–$y$ coordinate system is a straight line parallel to the $x$ axis (abscissa) and $a$ units of distance above sea level. Our distance $r$ from our origin 0 (the point where the $x$ and $y$ axes cross) increases steadily as we walk along this road in accordance with an algebraic formula involving $x$ and $y$ which we discuss later.

We now obtain different road profiles for our road by changing the functional relationship between $y$ and $x$. Instead of placing $y$ equal to a  constant, we allow it to change with $x$ in some way which may be simple or complicated. The simplest dependence of $y$ on $x$ is obtained by placing $y$ equal to $x$ ($y = x$) which is a linear or first-degree relationship (linear refers here to a straight line and the relationship is of the first degree as previously defined because $x$ appears to the first power).

We see at once that each point in our graph of this function on the $x$–$y$ plane is at the same distance from the $y$ axis (abscissa) as it is from the $x$ axis. All these points lie on a line which passes through the origin as we move along this road from the left of 0 to the right of 0. For every foot we move eastward to the right of 0 we move one foot vertically above sea level and for every foot we move westward to the left of 0 we drop one foot below sea level. At 0 on the road, which is quite a steep road, we are right at sea level.

Since a real road cannot be as steep as the functional equation $y = x$ requires it to be and still lie on the surface of the Earth, we picture the relationship between $y$ and $x$ as changing at some reasonable distances to the right and to the left of 0; the contour of the road may, of course, change as we move along it, bending either downward or upward. Since the contour of such a road is not described by a simple linear (straight line) equation (functional relationship) such as $y = x$, we picture our road as following this linear equation for only a small stretch (a small value of $x$) to the left and right of 0. The only information contained in this equation is the slope or pitch of the road in this small stretch. This is defined as the distance we move eastward or westward divided into either the rise above sea level or drop below it that we experience. In other words, the slope or pitch of the road when we are near our origin 0 is the change in our $y$ coordinate divided by the change in our $x$ coordinate. Since $y$ changes exactly by the same amount as $x$ does, we see that the slope of the road near 0 is exactly 1. We note that our distance $r$ from 0 when we are walking along such a straight road without any bends in it exactly equals the length of the road we have walked.

The pitch or slope of a surface is a very important concept which we encounter under various circumstances such as when we build the roof of a house so as to obtain a rapid runoff of rain or melting snow. The pitch is then specified as a definite rise of the roof per horizontal dimension of the house (e.g., 3 feet for a 15-foot horizontal stretch). We also deal with pitch or slope in constructing stairs or cement walks as well as in using screws, which are similar to spiral stairways. In climbing a spiral stairway, we rise a definite amount each time we go around the spiral once which gives us pitch. In climbing spiral stairways our motion may be pictured as walking in a horizontal circle while the circle moves upward at a fixed rate. In the same way, the pitch of the threads of a screw is defined as the distance the screw advances along its length for each complete turn of the screwdriver.

The simple equation $y = x$ is only one example of a straight road with a definite pitch or slope, in this case 1. We obtain a straight road with any slope by describing its contour with the equation $y = ax$ where $a$ is the slope and may have any numerical value we please. If $a$ is 2, the slope of the road in the neighborhood of 0 is 2; it rises 2 feet vertically for every 1 foot that it moves eastward. If $a = 1/3$, the slope is 1/3 so that the road rises 1 foot vertically for each 3 feet of displacement eastward. As $a$ increases, the road becomes steeper and steeper. If $a$ were infinite, the road would be a vertical precipice, rising vertically at 0 with no eastward displacement. The road would then be exactly perpendicular to the ground. If $a$ were finite but negative, as in the functional relationship $y = -3x$, the road would descend instead of ascend from west to east; to the left of 0 it would be above sea level and to the right of 0 it would be below sea level. During our lives we have all hiked along such roads and now we know how to describe them algebraically.

We have now described two different straight roads: one, $y = a$, which is always parallel to sea level at a height $a$ above

it, and a road $y = ax$ which is straight but rises steadily from
west to east with a slope $a$, penetrating the ground at sea
level only once at 0. Treated simply as algebraic equations,
$y = a$ and $y = ax$ differ in that $y = a$ has no roots—$y$ is never
zero, whereas the equation $y = ax$ has one root: $y$ is 0 when
$x$ is 0. We note also that the quantity $a$ in the equation $y = a$
is not a pure number because it must have the same physical
attributes as $y$ which is a distance; thus, $a$ must also be a
distance. In the equation $y = ax$, however, $a$ is a pure number
since $x$ itself is a distance so that $x$ contributes the distance
quality of the right-hand side of the equation to keep it in
step physically with the left-hand side. When we introduced
algebraic equations initially in the chapter on algebra, we
emphasized that if the quantities in the equations refer only
to numbers that have no physical counterparts, the quantities
on the two sides of the equations do not have to match in
terms of any physical measurements. We then say that the
terms in the equation are dimensionless or of 0 dimension.
But if we are dealing with a physical quantity such as a
distance as in the road example analyzed above, every term
in the equation describing the contour of the road must be a
distance. In the equation $y = ax$, $y$ is a distance (height above
sea level) and $x$ is a distance (the distance to the east of 0)
so that the equation is one dimensional (of the first power in
distance); thus, $a$ must be a pure number without dimensionality.

Having analyzed the two road contours $y = ax$ and
$y = b$, when $b$ is a constant, we now consider the road $y = ax
+ b$, where $a$ is a pure number but $b$ is a constant distance.
This contour is something like a combination of the two
contours $y = ax$ and $y = b$ and, indeed, we can obtain the
contour $ax + b$ by doing exactly what the right-hand side of
the equation tells us to do. To obtain the height of this road

above sea level at any distance $x$ from the origin 0, we add the height $b$ to the height of the road $ax$ at $x$. Starting at $x = 0$, we see that the height of the road above sea level is not 0 but $b$ (if $b$ is negative, the road is below sea level at $x = 0$). We see that if $b$ is positive, as we now assume, the road is at sea level when the value of $x$ is such that the right-hand side of the equation $y = ax + b$ is zero. This happens when $ax = -b$ or $x = -b/a$, which is at a distance $b/a$ to the left (to the west) of 0 if $a$ is also positive, as we assume. We now have all the information we need to construct the new road contour $y = ax + b$ from the old road contour $y = ax$. As a specific example we place $a = 1/2$ (a pure number, the slope) and place $b = 3$ feet (a distance). All distances are now to be expressed in feet. Placing $a = 1/2$ means that the old road rises 1 foot vertically for each 2 feet of eastward motion and $b = 3$ feet means that each point on the new road at the same eastward point $x$ as at the old road is 3 feet higher than the old road. We see that the road $y = (1/2)x + 3$ is exactly parallel to the road $y = (1/2)x$ but 3 feet higher than the latter road. In general, the road $y = ax + b$ is exactly parallel to the road $y = ax$ but $b$ feet higher than the latter road.

Treated merely as an algebraic equation, the functional relationship $y = ax + b$ is a statement that if $a$ and $b$ are known numbers, then for every number $x$ we can calculate a number $y$. By plotting all the number pairs $(x, y)$ on the $x$, $y$ plane defined by the $x$ and $y$ axes, we obtain points that lie on a straight line. We call $y = ax + b$ the equation of this line; a pure number $a$ is the slope of this line and $b$ is called its $y$ intercept (the height above sea level where it cuts the $y$ axis). Its $x$ intercept, a distance (its sea level point), is $x = -b/a$.

We can obtain any straight line we wish by an appropriate choice of the arbitrary constants $a$ and $b$ which tells us that a straight line is completely determined by two numbers: the slope of the line $a$ and its $y$ intercept $b$. This is equivalent to saying that two points determine one and only one straight line. To see how we can represent such a line, we consider the two points whose $(x, y)$ coordinates are $(x_1, y_1)$ for the first point and $(x_2, y_2)$ for the second point, where the subscripts 1 and 2 identify the points. Since the first point lies on the line, its coordinates $x_1$ and $y_1$ must obey the line equation $y = ax + b$ so that we have $y_1 = ax_1 + b$. This statement is also true for the coordinates $x_2$, $y_2$ of the second point so that $y_2 = ax_2 + b$. If we now subtract this equation from the first equation (we subtract the two left sides from each other and the two right sides from each other), we obtain the equation $(y_1 - y_2) = a(x_1 - x_2)$ since the $b$'s cancel each other. If we divide both sides of this equation by $(x_1 - x_2)$, we obtain $(y_1 - y_2)/(x_1 - x_2) = a$, which gives us the slope of the line determined by the two points. Our line equation now becomes $y = [(y_1 - y_2)/(x_1 - x_2)]x + b$. To find the constant $b$ we note that for $y = y_1$, $x$ must equal $x_1$ so we have $y_1 = [(y_1 - y_2)/(x_1 - x_2)]x_1 + b$ so that

$$b = y_1 - \left(\frac{y_1 - y_2}{x_1 - x_2}\right)x_1 = \frac{x_1 y_1 - x_2 y_1 - x_1 y_1 + y_2 x_1}{x_1 - x_2}$$

$$= \frac{-x_2 y_1 + y_2 x_1}{x_1 - x_2} = \frac{x_1 y_2 - y_1 x_2}{x_1 - x_2}$$

where to add $y$ to the term following it we use the common denominator $(x_1 - x_2)$. The equation of our line is then

$$y = \frac{y_1 - y_2}{x_1 - x_2} x + \frac{x_1 y_2 - y_1 x_2}{x_1 - x_2}$$

A simple numerical example will illustrate this equation. We consider the line passing through the point ($x_1 = 1$, $y_1 = 3$) and the point ($x_2 = 2$, $y_2 = 5$). Using the equation above for $y$ we replace $x_1, y_1, x_2, y_2$ by the numerical values just given and thus obtain the equation

$$y = [(3 - 5)/(1 - 2)]x + [(1)(5) - (3)(2)]/(1 - 2)$$

$$= (-2/-1)x + (5 - 6)/(1 - 2) = 2x + 1.$$

This is the equation of a line whose slope is 2 and which cuts the $y$ axis one unit of distance above the origin 0.

The motion of a body moving with constant speed along a straight line (for example, an automobile moving along a straight highway with its cruise control set at a definite speed) is a simple but important physical example of a linear (first degree) function which relates a distance to time. To show that we are talking about distance and time we write the equation that connects the two as $d = st$, where $d$ stands for distance and $t$ for time. If we plot $d$ against $t$ (distance against time), we obtain a straight line passing through the origin 0. In accordance with what we have said above, $s$ is the slope of the line and we see immediately that it stands for the speed of the body because $s$ must be a physical quantity which, when multiplied by time, it gives a distance; speed is precisely such a quantity so that the right-hand side of the equation $d = st$ is a distance like $d$. If the speed of the body is not constant, the curve that we plot is not a straight line. We discuss the physical significance of such a curve later.

We have discussed the graphical representation of the linear (first degree) equation that expresses a quantity $y$ as a function of another quantity $x$, where both $y$ and $x$, in this explicit example, are distances, to illustrate how graphing an equation gives us a quick insight into its basic mathematical properties. We shall see that this is also true for equations of degrees higher than one (quadratic, cubic, etc.).

How do we go from a first-degree equation to a higher-degree equation? Adding another $x$ term to a first-degree equation such as $bx$ does not do the trick because we then have $y = ax + bx = (a + b)x$ so that we still have a linear equation (first degree) with a slope equal to $a + b$ instead of $a$ so we get a steeper straight road if $b$ is positive. To obtain a road (a curve) that bends downward (a dip) or one that rises to a hill, we must add a term such as $x^2$ or $x^3$. If we have $y = ax + bx^2$ we have a quadratic functional relationship whose graph has a hill or a dip in it. To study the graph of such a quadratic we start with the simplest example $y = x^2$ and picture it as describing the contour of a road westward to the left of 0 and eastward to the right of 0. This contour has one feature in common with the linear function: the two roads are at sea level at the origin, because $x = 0$ there for both roads so that $y = 0$ also. But that is the only common feature because the contour $y = x^2$ is symmetrical with respect to the origin whereas $y = x$ does not have this symmetry. If we move 2 units of distance to the left of 0 (westward), as measured along the $x$ axis, so that our $x$ coordinate is $-2$, our height above sea level is $y = x^2 = (-2)^2 = 4$ units, since the square of a negative number is positive. If we now move 2 units to the right of 0 (eastward), as measured along the $x$ axis so that our $x$ coordinate is 2, we are then at the height $y = x^2 = (2)^2 = 4$ units above sea level. In other words, the curve that represents this simple quadratic function of $x$ is completely symmetrical with respect to the vertical $y$ axis (the ordinate). If we stand at 0, we are at sea level, the lowest point on the road, and we see that it rises steadily westward and eastward in exactly the same way. A road following the contour $y = x^2$ is not a real road because it would be at sea level only at one point $x = 0$ and then move off, moving upward forever on each side of 0. We remain near 0 for the present time and study the road's contours to the west and east of 0 but not too far from 0. We note first that the slope of the road near 0 is zero or close to it; the road there is very nearly parallel to sea level but rises with increasing slope as we move westward or eastward from 0. We note that the road tilts westward when we follow it to the west; we say that the slope is negative to the left of 0. To the right of 0 the slope tilts eastward; we call the slope positive to the right of 0. At the origin 0 itself the slope is zero. At the origin we are at the lowest point in the road; we call that a minimum (a smallest value) of the function $y = x^2$.

If, instead of following the equation $y = x^2$, the contour of the road followed $y = -x^2$, it would have the same symmetry as $y = x^2$, but owing to the minus sign, the slopes would be reversed and the road would have a hill at 0. We would then be at the highest point of the road (its maximum) which would be mathematically described as being the largest value of the function $y = -x^2$. The road $y = -x^2$ is obtained from the road $y = x^2$ by flipping this road around the $x$ axis, or reflecting it in a horizontal mirror placed at 0. We note from this example that wherever (at whatever value of $x$) the slope of the graph of a function is 0, the function has either a maximum or a minimum (either a hill or a dip in the road whose contour follows the function).

We now enlarge the quadratic function by adding a linear term to $ax^2$. Our quadratic becomes $ax^2 + bx$ and the function is $y = ax^2 + bx$ where $a$ and $b$ may be any real numbers. We see that the contour of the road this function describes is not symmetrical with respect to the $y$ axis (ordinate) of our coordinate system. To study the quadratic function $y = ax^2 + bx$ in more detail we consider the contributions to the total contour (as plotted in the $y$, $x$ plane) of $ax^2$ and $bx$ separately. We have already noted that the quadratic term $ax^2$ causes the slope of the contour (road) to increase eastward or westward if $a$ is positive (or to decrease if $a$ is negative). The term $bx$ does two things: (1) if $b$ is positive, the term increases the slope by the amount $b$ at each point to the east of 0 and decreases it for each point to the west of 0 and it does just the opposite if $b$ is negative; (2) it raises the contour by the additional amount $bx$ for each eastward point if $b$ is positive and lowers the contour by that amount for each westward point.

We now consider this quadratic form from a slightly different but informative point of view by noting that $x$ is a common factor in both of its terms so that we may factor $x$ out and, using parentheses, place it in front as a factor of both terms, thereby obtaining $x(ax + b)$. The quadratic equation then becomes $y = x(ax + b)$. Since a root of an equation is any value of $x$ which makes $y$ equal to zero, we see that this quadratic has the two roots $x = 0$, and the value of $x$ which makes the second factor of $y$ (the linear expression $ax + b$) equal to 0. We see at once that this expression is zero for $x = -b/a$. We thus have the two roots $x = 0$ and $x = -b/a$ so that the quadratic $y = ax^2 + bx$ describes the contour of an east–west road that is at sea level twice: at the point $x = 0$ (the origin) and at the point $-b/a$ units of distance to the east or west depending on the signs of $b$ and $a$. If $b$ and $a$ have the same sign (both negative or both positive), the fraction $b/a$ is positive so that $x = -b/a$ is negative and the road is at sea level (cuts the $x$ axis) $b/a$ units to the west of 0. But if $b$ is positive and $a$ is negative (or vice versa), the fraction $b/a$ is negative so that $x = b/a$ is positive and the road is at sea level again to the east of 0. In any case, we have two different roots, one at $x = 0$ and the other $b/a$ units

either to the east or west of 0, which means that the road has
a hill (a highest point) or a dip (a lowest point) halfway
between the sea level point 0 at the origin and the point $b/a$
to the west or east of 0.

We now point out another way of analyzing the contour
$y = ax^2 + bx = x(ax + b)$. This is a kind of product of two
straight lines: (1) the straight line $y = x$ which passes through
the origin 0 and each point of which is at equal distances
from the vertical $y$ axis (ordinate) through 0 and the horizontal $x$ axis (abscissa); and (2) the straight line $y = ax + b$
which passes through the point $b$ on the vertical ordinate
(the $y$ intercept) and the point $b/a$ on the $x$ axis (the abscissa).
To obtain the contour of the quadratic $y = ax^2 + bx$, we
multiply the height of every point on the straight line
$y = ax + b$ by the $x$ value of that point (its distance from the
$y$ axis). As a simple numerical example, we consider the
quadratic $y = 2x^2 + 3x = x(2x + 3)$. We obtain its graph
(contour) by multiplying the $y$ coordinate of each point on
the line $y = 2x + 3$ by its $x$ coordinate. Thus, the $y$ value of
the point $x = 3$ on this line is 9 [$y = 2x + 3 = (2 \times 3) + 3 = 9$].
We multiply 9 by 3 (the $x$ coordinate) to obtain 27 which is
the $y$ coordinate of the point on the quadratic contour $y$
$= 2x^2 + x$; its $x$ coordinate is still 3.

We complete our discussion of the complete quadratic
by adding the constant $c$ to the quadratic $y = ax^2 + bx$ to
obtain the standard quadratic equation $y = ax^2 + bx + c$
whose roots we discussed in detail in Chapter 3. Adding $c$ to
$ax^2 + bx$ simply means that the entire contour given by the
equation $y = ax^2 + bx$ is raised upward by the amount $c$ if $c$
is positive or lowered by this amount if $c$ is negative. We note
that this does not alter the shape of the contour but it may
change the two roots from real to complex; if $c$ and $a$ are
positive and $c$ is made large enough, the whole contour is

raised so high that the lowest point of the contour is above
sea level (above the $x$ axis) and the contour does not cut the
$x$ axis so that the roots are complex.

The contour (plot) in the $x$-$y$ plane described by the
quadratic equation $y = ax^2 + bx + c$ is that of a parabola
which is symmetrical with respect to some vertical axis if we
are describing a contour, such as that of an east–west road,
in a vertical plane. An excellent physical example of such a
contour produced by a moving body is the path of an object,
such as a javelin, thrown diagonally at sea level in the eastward direction. If the friction of the air against the javelin is
neglected and the thrower's position is taken as the origin 0,
the javelin leaves the thrower's hand with simultaneous vertical and horizontal motions which combine to produce a
parabola for the javelin's orbit. The symmetry axis of this
parabola is not the vertical $y$ axis (ordinate) through the
origin 0 but the vertical line through the point where the
javelin is at its greatest height above sea level. This point
along the horizontal (the $x$ axis) is midway between the
javelin's starting point $x = 0$, $y = 0$ (if we assume that the
thrower's height is 0) and the point where it hits the ground
again. The path of the javelin thus has two roots: $x = 0$ and
$x =$ the javelin's range. We give a more careful analysis of
the path of a javelin in a later chapter.

We have limited our discussion here to parabolas that
have vertical symmetry, but we can certainly draw parabolas
that are symmetrical with respect to any line which cuts our
vertical ordinate and horizontal abscissa. Such a parabola
is not described by the standard quadratic equation
$y = ax^2 + bx + c$. Since neither the vertical $y$ axis nor the
horizontal $x$ axis is a line of symmetry for a parabola tilted
in any way in the vertical plane, the equation that describes
such a parabola must involve $x$ and $y$ in the same way and

not show any bias toward one or the other. On the left-hand side of the standard equation $y$ appears to the first power whereas on the right-hand side $x$ appears both as a square to the second power and as a first-power term. If we write the standard quadratic in the form $y - ax^2 - bx = c$, we see at once the asymmetry between $y$ and $x$. To obtain an equation that gives the contour of a parabola that may have any orientation in the vertical plane we must include a $y^2$ term and an $xy$ term. We then obtain the general quadratic in $x$ and $y$, viz., $ay^2 + bxy + cx^2 + dx + ey = f$, where $a$, $b$, $c$, $d$, $e$, $f$, are arbitrary real numbers that may be chosen any way we please.

By choosing these six constants appropriately we obtain a series of different contours or curves which, as a group, are called conic sections. The reason for this name is that if one slices through a perfect vertical cone (let us say a cone of cheese) standing with its base on the level ground, the contour of the cut is described by this general quadratic equation. This contour may be a circle, an ellipse, a parabola, or a hyperbola. We come back to conic sections in a later chapter but note here that if $a$ and $c$ are chosen equal to 1 and $b$, $d$, and $e$ are placed equal to 0, we obtain $y^2 + x^2 = f$, which is the equation of a circle whose center is at the origin 0 and whose radius is $\sqrt{f}$; $f$, of course, must be positive. If $a$ and $c$ are unequal, positive numbers, $b$, $d$, and $e$ are 0, and $f$ is a positive number, the quadratic obtained $(ay^2 + cx^2 = f)$ describes an ellipse which is interesting because the planets move around the sun in ellipses, as first discovered by Johannes Kepler by trial and error and as later deduced from gravitational theory by Isaac Newton.

Thus far we have described a road whose profile or contour has either one dip or one hill in it, rising on either side (east and west) of the dip or falling (east and west) on either side of the hill. We now leave such a parabolic road and consider the algebraic equation that describes the contour of a road that has one hill and one dip. Such a road cuts the sea level three times: once when it falls below sea level to form its dip, the second time when it rises above sea level to form the hill, and the third time when it falls off from the hill and penetrates the sea level again. This means that the equation describing this contour must have three roots, which means it must be a cubic, such as $y = ax^3 + bx^2 + cx + d$, where $a$, $b$, $c$, and $d$ are real numbers. We thus arrive at a very important conclusion—for each additional hill or dip in the contour the equation describing the contour must have an additional term in it with $x$ raised to a power larger by 1 than the highest power of $x$ in the equation describing the contour with one fewer hill or dip.

Thus, the equation for the contour of a road with $n$ hills and dips (if $n$ is even, the number of hills must equal the number of dips, each equal to half of $n$; if $n$ is odd, the number of hills and the number of dips differ by one) is of the $n$th degree in $x$. This means that the highest power of $x$ in the equation is $x^n$. The general equation for any contour is thus written as $y = ax^n + bx^{n-1} + \cdots + cx + d$, where the coefficients $a$, $b$, . . . , $c$, $d$ are real numbers, and the dots stand for terms that we have not written down but are understood.

We complete this chapter on the graphs of algebraic equations with a discussion of simultaneous algebraic equations, which are met in many problems in physics, chemistry, engineering, economics, etc. Here we consider only the simplest type (a pair) of simultaneous equations—those involving just two unknowns, which must be determined together by solving the two simultaneous equations. In the algebraic equations we have already discussed such as the quadratic equation $ax^2 + bx + c = 0$, only one unknown is present and solving the equation means finding the roots of the equation. We have seen that we can obtain the roots by an algebraic procedure or by plotting the quadratic polynomial $y = ax^2 + bx + c$ and noting where the plotted curve in the $x$-$y$ plane cuts the $x$ axis. We can solve simple simultaneous equations having two unknowns $x$ and $y$ either algebraically or graphically.

If the two simultaneous equations are $ay + bx = c$ and $dy + ex = f$, where $a$, $b$, $c$, $d$, $e$, and $f$ are real numbers, we can solve these two equations for $x$ and $y$ algebraically by multiplying the entire first equation by the number $d$ and the second equation by the number $a$, and then subtracting one of these multiplied equations from the other. The two multiplied equations are $day + dbx = dc$ and $ady + aex = af$ with the first term on the left-hand side of each equation the same so that this term drops out on subtracting the second equation from the first, leaving the single equation $dbx - aex = dc - af$ for $x$. Since $x$ is a common factor of each term on the left-hand side, we may write this equation as $x(db - ae) = dc - af$. On dividing both sides by $(db - ae)$ we obtain the solution of the simultaneous equations for $x$, viz., $x = (dc - af)/(db - ae)$. On substituting the given numbers for $a$, $b$, $c$, $d$, $e$ in the fraction on the right-hand side, we obtain the desired numerical value for $x$, and on substituting this number for $x$ in either of the two simultaneous equations we obtain a single equation for $y$ expressed entirely in terms of the numbers $a$, $b$, $c$, or $d$, $e$, $f$ so that the numerical value of $y$ is obtained. The two simultaneous equations for $x$ and $y$ are thus solved algebraically.

We can also solve these equations graphically by plotting the contour described by each equation in the $x$-$y$ plane. Rewriting the two equations in the form $y = c/a - (b/a)x$ and $y = f/d - (e/d)x$, we see that each of these equations is the equation of a straight line. The solution of these two equations is then given by the $x$, $y$ coordinate of the point in the $x$-$y$ plane where the two lines intersect.

The use of graphs to elucidate algebra and to show its relationship to geometry was first introduced by the great French philosopher and mathematician René Descartes. The framework of the graph with its ordinate and abscissa as the two perpendicular reference axes, called a Cartesian coordinate system, is used extensively in all branches of mathematics and physics. As with many great discoveries in the sciences, Descartes did not consciously formulate the concept of the Cartesian coordinate system (the $x$-$y$ plane) based on his desire to represent graphically the position of a point. Instead he stumbled upon the idea as he lay in bed one day while staring out a window. He noticed that a branch outside the windowpane moved to the left and right and up and down across the panes as the wind blew through the tree. Descartes noticed that he could describe the position of the branch as three windowpanes to the right and two windowpanes up from the side of the window or by any other combination of horizontal and vertical coordinates. It was a simple matter to transfer the idea to a conceptual $x$-$y$ plane and substitute points for the windowpanes.

Introducing a coordinate system to locate points on a plane is equivalent to laying off a mesh or a grid on the plane which is produced by the intersections of a set of lines parallel to the ordinate and a set of lines parallel to the abscissa. If the two sets of lines are exactly perpendicular to each other (the simplest kind of coordinate system), the mesh or grid looks like a checkerboard so that the plane is covered with small squares. If, however, the two sets of lines are oblique to each other, the mesh consists of diamonds. Later we shall discuss other types of coordinate systems and see that the choice of the proper coordinate system can greatly simplify a problem and lead to its quick solution.

# CHAPTER 5

# The Geometry of Straight Line Figures

*One cannot escape the feeling that these mathematical formulae have an independent existence and an intelligence of their own, that they are wiser than we are, wiser even than their discoverers, that we get more out of them than was originally put into them.*
—HEINRICH HERTZ

We have seen that the step from arithmetic to algebra is merely the transference of the rules of numbers to letters, which, in a sense, makes algebra more abstract than arithmetic. But this difference is more apparent than real because numbers are abstractions also unless we associate them with measurable entities. Since a measurement is never precise, however, the number assigned to such a measurement is something of an abstraction and must be taken on faith. We accepted this abstraction when we introduced the ordinal aspect of numbers by assigning integers or fractions to points on a line because the points themselves are abstractions. Numbers become abstract even when we assign them to collections of discrete entities such as people or atoms. We can certainly speak of 5 people or the population of a town but things are not so clear when we are dealing with atomic entities. To say that we have one atom or one electron in a box is an abstraction since we can never isolate an atom or an electron with a certainty that permits us to speak of only one. We also run into arithmetic abstractions (uncertainties) when we deal with very large numbers so that to speak of a billion people or a trillion dollars or the number of hydrogen molecules in 2 grams (one mole) of hydrogen is an abstraction. As long as we deal with numbers as symbols, which obey certain rules (addition, multiplication, etc.) we may consider them as precise entities and arithmetic becomes a game involving these entities or symbols.

Since algebra is just a carryover of arithmetic to letters (symbols different from the arithmetic numbers) which may represent quantities of entities that need never be specified, even in a numerical sense, it is more abstract than arithmetic. But in working with geometry we have to give up some of the abstraction that delights us in working with algebra because geometry deals with the spatial universe which is governed by geometrical laws. From that point of view, geometry is part of physics, which is the basic science of the universe. Geometry therefore enjoys a far greater reality than does algebra. But in using geometry we shall employ algebra and arithmetic to formulate our concepts precisely. The abstraction or uncertainty that we encounter in geometry is that the geometry that seems to describe our world that we accept when we begin studying geometry in school may not correctly describe the space of the entire universe or the space inside or on the surface of a very massive compact sphere. We encounter geometric uncertainty also when we penetrate into the internal space of a proton, an electron, or a neutron.

If only one kind of geometry existed and that were the geometry we studied in school (called plane or Euclidean geometry), the universe would be governed by that geometry and much of the abstraction associated with geometry would disappear. For many years Euclidean geometry was thought to be true and philosophers, mathematicians, and scientists held up Euclidean geometry as the purest of truths never to be questioned. But we now know that Euclidean geometry is a false prophet whose dogma applies to no part of the universe and, indeed, can never be demonstrated as true. That being so, we must consider just how the study of geometry arose as well as its intellectual foundations. Geometry arose much earlier and developed much more rapidly than algebra because it was essential for the technological growth of civilization. Geometry was required whether people were building houses, laying down roads, digging wells, or establishing property rights by surveying land. The early Greek, Egyptian, and Roman geometers were manifestly brilliant in their complete mastery of geometry and in their application of the principles of geometry to the solution of all kinds of practical problems. The magnificent structures, with their remarkable symmetry and great strength, produced by these early civilizations are the evidence of this mastery. Navigators, geographers, map makers, and astronomers had to be good geometers to pursue their vocations. Whereas algebra was more or less an intellectual game to ancient civilizations, knowledge of geometry was essential and so the books of Euclid and the treatises of Pythagoras were widely studied.

Before considering the elements of Euclidean (plane) geometry we discuss certain general features that are common to all geometries, whether they conform to the real space of our universe or some imaginative or fictitious space that we invent. We may, of course, be as fanciful as we please in constructing or inventing a purely theoretical geometry but ultimately we can be satisfied only if the geometry describes the spatial relationships of our universe. Pythagoras, Euclid, and Archimedes believed that their geometry performed precisely that function; this belief persisted into the 19th century. Kepler, Galileo, and Newton developed their physics (laws of planetary motion, laws of the motions of bodies in general, etc.) in accordance with Euclidean geometry whose truths seemed self-evident. But they really had no choice since only Euclidean geometry was known to them and nobody dreamed that another kind of geometry could exist. Ever so perceptive and profound a philosopher as Immanuel Kant argued that Euclidean geometry is the one universal, God-given truth that cannot be questioned.

The big break with Euclidean geometry occurred in the mid-19th century when the great German mathematician Karl Friedrich Gauss began to develop the geometry of curved surfaces. Clearly the geometry of figures drawn on the surface of a sphere cannot be plane (Euclidean) geometry since the surface of a sphere is not a plane nor can it be made to conform to a plane. Straight lines cannot be drawn on such a surface and Euclidean geometry deals with straight lines as commonly understood. Gauss's non-Euclidean geometry of curved surfaces was readily accepted as a new geometry and raised no particular criticism as to its validity because it was quite clear to mathematicians of the time that the properties of the figures drawn on a curved surface can be treated by a spherical geometry designed for such a surface or by Euclidean geometry applied to the same figures without any reference to the curved surface. To be specific we consider a great circle on the surface of the Earth that connects the north and the south poles. Walking along such a surface, we appear to be walking along a straight line, whereas to a person observing us from outer space we are walking along the arc of a circle which is just an ordinary Euclidean circle. Thus, we have the choice of using a special kind of surface geometry, as Gauss did, to describe geometric figures on any kind of curved surface embedded in space, or using Euclidean geometry to describe those same figures as existing in space in their own right without reference to a curved surface. This is all well and good if the geometry of all of space itself is Euclidean, which every sensible, commonsensible person accepted as self-evident, because to think otherwise would have meant (and still means) that space is curved in some way. But this statement appears to be contradictory: How can all of space be curved if there is nothing outside the universe (where the universe is defined as containing everything) into which it can bend or curve? We shall see later, however, that this concept of "curved" is too restrictive and the "bending" or "curvature" of space can be defined in such a way that it can actually be measured so that determining the geometry of space, as a physical feature of space, becomes a part of physics. Indeed, Gauss himself, who was almost as good a physicist as he was a mathematician, suggested that astronomers try to determine whether or not a huge triangle in space, formed by the three lines connecting each of three stars to its two neighbors, is governed by Euclidean geometry.

Though this forbidding astronomical adventure was never pursued, two great mathematicians, the Russian N. I. Lobachevsky and the German G. F. Riemann (Gauss's greatest student) developed two different non-Euclidean geometries of space: Lobachevskian (hyperbolic) geometry and Riemannian (elliptical) geometry. Indeed, these geometers went beyond the geometry of physical space, which is defined as three-dimensional, and developed their two non-Euclidean geometries for any number of dimensions, which play very important roles in Einstein's general theory of relativity, perhaps the single greatest creation of the human mind.

To understand the origin of different geometries we first consider the basic elements of all geometries and then see how differences in these basic elements lead from one geometry to another. These basic elements are of a threefold nature. First, certain undefinables are introduced which can be combined to produce derived quantities; these derived quantities are then defined in terms of the undefinables. This process is similar to the way a dictionary is constructed. The editor of a dictionary starts with basic words, whose meanings are accepted without definition; the meanings of all other words in the dictionary are then expressed in terms of the undefined words or in terms of words that have already been defined plus the undefined words.

Since all geometries deal with space and space consists of points, we accept the point as an undefinable. We may, of course, identify a point, as we did in arithmetic, by means of a symbol (a number) but that does not define a point. Since geometry also deals with distances between points, we must accept distance as an undefinable, which we conceptualize as we do a point. We shall see later that we can perform a series of physical operations (measurements) which give us what we call a distance, but these operations do not define distance. From the concepts of a point and of distance we now define a straight line as the shortest distance between any two points. The concept of "shortest" in this definition of a straight line still has to be explained, but we defer that discussion for the moment.

With this discussion of undefinable, definable, and derived quantities, we now consider the second group of elements on which every geometry rests: the axioms, which, in elementary geometry textbooks, are defined as "self-evident truths." As we shall see, this characterization of the axioms of geometry is unacceptable without further study. The axioms are a set of statements which, in part, are the axioms of logic, of algebra, and of arithmetic and, in part, statements about the nature of space which we accept as true. The greatest mathematician of the 20th century, David Hilbert, carefully analyzed these axioms and showed that they must be governed by certain criteria if they are to lead to a complete geometry.

The statement that "two things equal to the same thing are equal to each other" is an example of an axiom of a logical nature. We can accept this axiom without questioning it because it is a self-evident truth. The algebraic and arithmetic axioms are simply the rules of addition and multiplication that define arithmetic and algebra as developed in the foregoing chapters. In other words, we carry over to geometry all of arithmetic and algebra. Geometry, in a sense, is thus the algebra and arithmetic of geometric figures. Thus, we may add one kind of figure to another (or subtract one from another) to obtain a third kind of figure and we may multiply lengths of lines together to obtain a new kind of geometrical entity.

We now state two axioms which are of a spatial nature and are the basis of Euclidean geometry: (1) one and only one straight line can pass through two different points; and (2) given a straight line and a point outside this line, only one line parallel to the given line can pass through the point. This so-called "parallel axiom" is Euclid's fifth axiom which is perhaps the most famous statement in the history of geometry. The profound criticisms and ultimate denial of this axiom led Gauss, Lobachevsky, and Riemann to derive their non-Euclidean geometries.

The axioms of geometry must fulfill certain basic requirements before a complete self-consistent geometry can be built on them. The most important of these are the following: (1) no two axioms must contradict each other (the axioms must be self-consistent); (2) the axioms must not be redundant (no two axioms must say the same thing); and (3) the axioms must be complete so that all possible theorems that are consistent with the geometry can be proved.

From these considerations we see that a geometry is a vast intellectual enterprise consisting of undefinables, definables, axioms, and theorems. Since the number of theorems is vast, indeed endless, geometry itself has no limits. Considering that geometry is also part of all phases of science and may be utilized to deal with the unfolding of events in space and time, we see that geometry is the richest of all branches of mathematics. Unfortunately, the beauty and intellectual wonders that geometry offers are never revealed to high school students, who must study plane (Euclidean) geometry, and so they look upon geometry as a tedious enterprise that they must endure as part of their education.

Before discussing some of the basic theorems in plane geometry, we introduce another concept which we did not deal with in either arithmetic or algebra: the concept of angle. We introduced arithmetic by considering points on a line and then labeling equally spaced discrete points by ten symbols (the numerals) which defined the integers of our number system, and distances along this line from some point 0 (our origin) which we pictured as measured by motion (steps) along the line. But we cannot develop geometry in terms of a single line; we introduce other lines which we may pass through the origin 0. We consider one such line and introduce a method or procedure for identifying it which distinguishes it from any other line through 0. To do this we first picture our 0X line, that is, our $x$ axis (abscissa), as pointing to some fixed object in space such as, for example, the star Vega. If we stand at 0 and look along the $x$ axis, we will see Vega. We now turn to the left until we are looking along the line we wish to identify. The amount of turning we must do from the direction of the $x$ axis to the direction of the given line is called the angle at 0 formed by the intersection of the $x$ axis and the given line. In other words, angle measures turning or rotation just as distance measures motion from point to point on a line. Note that distance (motion along a line) is measured in distance units such as centimeters (cm) or kilometers (km) whereas an angle is a pure number and has no spatial dimensions.

Although angle has no spatial dimensions, it can be measured in units which play the same role for angles as the centimeter or inch does for distances. We again look directly along the $x$ axis (abscissa) and now turn completely around until we are again looking along the $x$ axis. We divide this complete turn (rotation) into 360 equal units of turning (equal angles) and call each of these units of angle one degree (1°). Thus, a complete rotation contains 360°. If we divide 1° into 60 equal parts, we call each of these parts one minute (1') of turning (an angle of 1'; 1° therefore equals 60').

The minute is used in defining the resolving power of the normal human eye. To explain this concept we consider two fixed dots on a screen viewed from a certain distance by the normal human eye. The two rays of light entering the eye (one ray from each dot) intersect at the eye at an angle which depends on the actual distance between the two dots measured on the screen (their actual separation) and the distance of the eye from the screen. The larger the separation and the smaller the eye-to-screen distance, the larger is this angle which is called the apparent separation. This angle is just the amount the eye has to rotate from one dot to the other to focus each dot in turn at the center of the eye's retina. As the eye approaches the screen, the angle through which the eye has to turn (the apparent dot separation) increases, but as the eye recedes from the screen the dots appear to approach each other until they appear to merge into a single dot (the angle between the two rays is zero). If the angle between the two rays (the angle through which the eye must turn) is larger than or equal to 1′ (1/60th of a degree) the normal eye sees the two dots as distinct (it resolves the impression on the retina into two images) but if this angle is smaller than 1′, the two dots merge into one. We therefore say that the resolving power of the normal human eye is an angle of 1′. We note that whenever we look at objects at various distances from us, our eyes register only their apparent—not their true—separation; our brains then call upon their vast memory banks which contain the kind of information and correlations that enable us to estimate the true separations. We are never aware that this process is happening because the brain integrates all the information so rapidly that we are given the true separations of the objects directly by this integration process.

We describe another important deduction about our external environment that stems from the rotation of our eyes. If we had only one eye, our depth or distance perception would be greatly impaired, but two eyes produce a stereo-scopic effect which enables us to estimate the distances of objects that are not too far away. Our two eyes do this by rotating laterally toward each other. The greater the distance an object is from our eyes, the smaller is this turning. As the object approaches us or as we approach it, our eyes turn inward and the brain registers this fact as a decrease in distance.

In many natural phenomena, angular measurements much smaller than 1′ (or one minute of arc, as it is called) must be measured. For that reason the angular minute is divided into 60 equal subdivisions, each of which is called a second of arc written as $1''$ so that $1° = 3600''$. One second of arc is an exceedingly small angle which we may describe fancifully as follows: Balance a dime on its rim at a distance of $2\frac{1}{2}$ miles from your eye and follow a tiny bug with your eye as it moves diametrically across the dime. Your eye will turn through an angle of $1''$ of arc as it follows the bug; if the bug were viewed through a modern astronomical telescope, with all its electronic devices for accurate measurements, the viewer could accurately tick off the bug's position on the dime for each one hundredth of the distance across the coin. In other words, modern telescopes can measure the change in the direction of a moving object (a star or a planet, for example) with an accuracy of one hundredth of a second of arc.

Because of their great angle-measuring accuracy, telescopes have meant the difference between a very limited naked-eye Earth-based astronomy and the vast cosmic enterprise that has revealed the expanding universe, the structure, motion and chemistry of stars and planets, and the structures and dynamics of galaxies. This great adventure began in 1838 when the German astronomer and mathematician Friedrich W. Bessel first calculated the distance of a star (61 Cygni) by measuring its stellar parallax, which is defined as half the angle through which the line from the Earth to the star rotates when the Earth in its orbit goes halfway around the sun. The stellar parallax (always written as $p''$ and measured in seconds of arc) is half the angle through which a star's position appears to shift every six months as the Earth moves from any point on its solar orbit to a diametrically opposite point. Thus, as the Earth shifts back and forth by 186,000,000 miles from one side of the sun to the other, all the stars appear to shift back and forth by angular amounts that depend on their true distances from the sun; the more distant a star, the smaller the parallactic shift. Thus, the parallax of the nearest star, Alpha Centauri, is 0.752″ (slightly more than 3/4 second of arc). We may consider this idea from two other points of view. First, picture yourself on a star following the Earth in its orbit; half the angle through which the line of sight from your eye to the Earth rotates is the star's parallax. If you were on Alpha Centauri, the Earth's orbit would appear to be about $1\frac{1}{2}$ times as large as a dime viewed from a distance of about $2\frac{1}{2}$ miles. Second, we may describe the parallax of a star by picturing a giant, with his two eyes separated by 186,000,000 miles, viewing a star first with one eye and then the other and noting that, when so viewed alternately, the star appears to shift back and forth against the distant background stars; half of this angular shift is the star's parallax. This is exactly like the apparent shift of an object one observes if the object, held at a definite distance from one's eyes, is viewed first with one eye and then the other. The greater the distance of the object, the smaller is the apparent shift. In place of a giant, we use a telescope fixed on the Earth, which is shifted by 186,000,000 miles every six months so that the telescope is equivalent to the two eyes of the giant if the giant alternates from one eye to the other every six months. The astronomer measures the semiannual parallactic shift by using the telescope like a very large camera and photographing the star at six-month intervals and then comparing the images on the photographic plates. The six-month angular shift of the star is then obtained by first measuring the actual shift, in distance units, of the image of the star on two successive photographic plates (six months apart) and then dividing this distance by twice the length (in the same distance units) of the telescope and then multiplying this quotient by the factor 206,265. Thus, if the two photographic images of a star, taken six months apart, by a telescope with a length of 1000 centimeters, are separated by 2/1000th of a centimeter, the parallax of the star is 0.2″ (two tenths or one fifth of a second of arc). Note that if we lived on Jupiter, which is five times as far from the sun as the Earth, all parallaxes of stars would be five times larger.

From its measured parallax a star's distance can be computed arithmetically in miles by multiplying 206,265 by 93,000,000 and dividing this product by the star's parallax in seconds, where 93,000,000 is the sun's mean distance from the Earth. We express this quantity algebraically as the general formula $d = (206{,}265)A/p''$ where $d$ is the star's distance, $A$ is the mean distance of the Earth from the sun, and $p''$ is the star's parallax. Note that the distance $d$ is expressed in exactly the same units of distance as is $A$. If $A$ is given in miles, then $d$ is also given in miles. As an actual stellar example we find that the distance of the closest star (Alpha Centauri) in miles is $(206{,}265)(93{,}000{,}000)/(3/4)$ which equals 26.5 trillion miles. We have gone to such great lengths in our discussion of angle measurements to show how such a simple concept as angle and its accurate measurements permit us to calculate the single most useful number that we can assign to a star. The great importance of this number for astronomy is that astronomers can use it to translate apparent features of stars to actual characteristics. From the brightness of a star (the way a star looks to us) and its distance we can thus calculate the star's luminosity (the total amount of light it emits every second in all directions). We can also apply angle measurements to the measurements of the sizes of such solar system objects as the sun, moon, and planets. The sun and moon appear equally large to us, as demonstrated by the almost exact coverage of the sun's disk by the moon during a total solar eclipse. When we allow our eyes to go from one edge of the full moon's disk to the diametrically opposite edge, our eyes turn through half a degree. This is also true for the sun's disk. We thus say that the apparent sizes (diameters) of the sun and moon are 1/2 degree, from which, and from the knowledge of their distances, we can calculate their true diameters (true sizes).

So important is the angle concept in mathematics, science (particularly physics), and even life processes, that we discuss some other interesting examples before passing on to the next topic. In nature we find many remarkable phenomena and structures in which precise relationships among angles are very important. Thus, in crystals (e.g., diamonds, salt, sugar, ice, various ores and precious stones) the atoms or molecules arrange themselves in precise lattices, with the lines from any atom or molecule to its nearest neighbors intersecting each other in angles that are exactly the same no matter which atom or molecule in the crystal we consider. A very beautiful example of this phenomenon is the snowflake which always has a hexagonal (six-sided) form with each of the radiating spokes of the flake making an angle of exactly 60° at the center with each of its two neighbors. Other such examples are the angular arrangements of the threads of spiderwebs and the exact hexagonal shape of a honeycomb cell.

Although we are not consciously aware of our mental processes when we differentiate between a large and a small object, we do so, not because our eyes measure differences in distances (sizes) and then send this information to the brain, but because the rays of light from the top and bottom of the large object meet at a larger angle in the eye lens than the similar rays do from the small object. The eye thus measures angle differences when these various rays impinge on the retina and the brain then converts these angle differences to size differences. Another important example of this process is the magnification power of a pair of binoculars or a telescope. When we look at a distant object with our naked eyes, the rays from the top and bottom of the object intersect at a small angle at the eye, but when these rays pass through binoculars or a telescope before entering the eye, they are tilted away from each other by the binoculars or telescope so that to the eye looking through the telescope the image produced by the telescope appears larger than the naked-eye image.

This discussion brings us to the very important use of angle measurements in studying the way rays of light behave when they pass from one medium to another such as from a vacuum into air, or from air into glass or water, or from one kind of glass into another as in a lens system. In a vacuum (empty space), light waves (radio waves, infrared, ultraviolet, X rays, etc.) travel in straight lines and with exactly the same speed, but when light enters a material medium such as glass, it slows down, with different colors of light traveling at different speeds. The direction of the ray (the direction of its path) changes when the ray passes from one medium to a different one because its speed changes. This change in direction of a ray of light is called refraction and the angle which measures this change in the ray's direction is called the angle of refraction.

To give what we have just said a deeper physical meaning we consider a ray of light striking a smooth, flat glass surface obliquely. At the point where the ray enters the glass we erect a line perpendicular to the surface (it makes an angle of 90° with any line on the surface); the angle the ray makes with this perpendicular line (also called the normal to the surface) at the point where the ray and the perpendicular line meet is called the angle of incidence of the ray. The reflected part of the ray (some of the ray is always reflected from any surface it strikes) leaves the surface at an angle to the normal line (called the angle of reflection) exactly equal to the angle of incidence. Most of the ray, however, enters the glass, but its direction changes; it is tilted or rotated toward the normal so that the angle (called the angle of refraction) it makes with the normal is smaller than the angle of incidence. This angle is different for different kinds of glass (different media) and each color (red, green, blue, etc.) has its own angle of refraction, with red having the smallest (its path is bent least toward the normal) and with violet having the largest. If the smooth surface is round (the surface of a lens) instead of flat, the perpendicular at each of its points differs in direction from point to point so that the angles of incidence of a bundle of parallel rays and, hence, the angles of refraction of these rays, are different. A lens, therefore, with two round surfaces such as the lenses in our eyes, cannot bend all the rays that strike it in such a way as to bring them together at a single point to produce a sharp image. A good camera lens therefore consists of an array of lenses of different glass types (with differently curved surfaces) cemented together. With the right kind of glass for each lens and with the proper roundness of each surface of each lens, such a cemented array of lenses can produce very sharp images.

We have discussed the angle concept in some detail to emphasize its importance in our daily lives as well as in geometry. We now go on to develop the basic elements of geometry in all of which the angle concept plays a very crucial role. In our development of arithmetic we started with points on a line, but a single line cannot lead to geometry. If we now introduce an infinitude of such lines densely packed next to each other, without intersecting each other, they form a plane. Such lines are to a plane as the points on a line are to a line. In other words, just as a line is the collection of all the points on it, so a plane is the collection of all parallel lines on it. If we limit ourselves to a line, we can move in only one direction—to the right or left along the line. The domain of points on the line is therefore a one-dimensional manifold, so that we cannot deduce anything about the nature of geometry from arithmetic alone. In going from points on a line to points on a plane we go from a one-dimensional to a two-dimensional manifold on which we can construct a geometry—plane geometry or the Euclidean geometry of a plane. Since space itself can be constructed by stacking an infinite array of closely packed parallel planes next to each other, we obtain, in that way, the domain of all the points in space which is a three-dimensional manifold governed by, or described by, Euclidean geometry. We sum this up by saying that points in a line describe a one-dimensional space or manifold, that lines in a plane describe a two-dimensional space, and that all the planes we can construct describe our three-dimensional space. This is illustrated physically in nature by the arrangement of atoms in a cubic crystal (a cubic lattice) such as that of salt. In a salt crystal the sodium atoms and chlorine atoms are arranged alternately and equally spaced along parallel lines, which lie in equally spaced parallel planes. To observe this lattice structure directly, we would have to reduce our size to one hundredth of a millionth of a centimeter so that we could creep into a salt crystal between its atoms.

Although we cannot physically explore a salt crystal, we can send in messengers that can penetrate into crystals and return to us to reveal exactly the crystal lattice structures of any kind of crystal. These messengers are either X rays or very rapidly moving electrons. X rays are very high-frequency (very rapidly oscillating) bundles of electromagnetic energy (photons of extremely short wavelength) which are no larger than the spacing distances of the atoms in a crystal. Such X rays (photons) are not reflected from an ordinary smooth surface like visible photons, but they are reflected from the successive layers (planes) of atoms in crystals and when they impinge on a photographic plate they produce a pattern (called a diffraction pattern) from which an accurate picture of the lattice structure can be constructed.

Returning now to the plane, we note, as already stated, that we can cover a plane or define it by an array of closely packed parallel lines but we proceed here somewhat differently because sets of parallel lines can include lines that do not lie in the same plane. To avoid any ambiguity we consider sets of intersecting lines and note that any two intersecting lines determine or define a single plane. We label the point where these two lines intersect with the numeral 0 and call 0 the origin of our plane as defined by the two intersecting lines. We call one line the X line ($x$ axis or abscissa) and lay it off along the east–west direction and call the other line the Y line ($y$ axis or ordinate) which lies along some northeast–southwest direction as specified by the angle at which these lines intersect at 0. Once we have chosen these two lines (specified their angle of intersection), we can locate any point on the plane by stating how far we must walk along the $x$ axis and then how far we must walk parallel to the $y$ axis to reach the point. We can thus reach any point on the plane by walking two definite distances along two different intersecting straight lines. These two distances are called the X and Y coordinates of the point in our frame of reference (coordinate system), that is, in our two sets of intersecting parallel lines.

We may make the angle of intersection of the 0X and 0Y lines at 0 as large as we please, but one special angle greatly simplifies our geometry. We note first that if the angle of intersection is smaller than 90°, the two directions 0X and

0Y are not independent of each other. Indeed, in taking the direction of 0Y as northeast or southwest we imply that in walking along 0Y we are also moving along the 0X direction (eastward). As we increase the angle of intersection of 0X and 0Y at 0 by turning the 0Y line northward, we reduce our eastward motion or displacement along 0X when we move along 0Y until, for an intersection angle of 90° (0Y is then due north–south), the 0Y and 0X directions are independent of each other. This means that moving along 0Y does not change our position along 0X; we experience no east–west motion. A coordinate system of parallel east–west lines intersecting parallel north–south lines is called an orthogonal Cartesian coordinate system.

Using this orthogonal coordinate system we can now define more precisely the concept of dimensionality. We spoke of the points on a line as constituting a one-dimensional manifold. This means that once we have chosen the origin 0 on the line the position of any point on the line is given by a single number (positive or negative), which is the distance of the point to the right or left of 0. But for a point on our plane we must give two (and only two) numbers to locate it: its distance eastward or westward from 0 and its distance north or south of 0. Hence the plane is two-dimensional. We also note that a plane has only two independent (perpendicular) directions, which may be chosen arbitrarily. Any other direction on the plane can then be represented as a definite combination of these two independent directions.

As we have noted, the geometry on a plane (a flat surface) is Euclidean so that it stems from Euclid's axioms, the most important of which is the parallel axiom which states the following: Given a line on a plane and a point off the line, there is only one line that can be drawn through the point parallel to the given line. Accepting this statement and

Euclid's other axioms, which are essentially of a logical nature, we can apply the rules of arithmetic and algebra to deduce the theorems of plane geometry. Before we discuss these rules, however, we introduce the concept of area which is obtained by multiplying distances, to which we are led by considering points on the $x$-$y$ plane defined by the $x$ axis (abscissa) and the $y$ axis (ordinate). Each such point P defines a distance from the origin 0 to the point itself but it also defines or represents an area which we describe below. The point P itself is identified by its X coordinate (its distance from the $x$ axis measured along a line parallel to the $y$ axis) and its Y coordinate (its distance from the $y$ axis measured along a line parallel to the $x$ axis). Thus, the point $(3, 4)$ is 3 units of distance (centimeters, for example) to the east of the $y$ axis and 4 units of distance north of the $x$ axis. The distance of the point P from 0 along a straight line from 0 to P written as 0P is always smaller than the sum of the X and Y coordinates of P. In the present example, $0P < 3 + 4$ where $<$ means that the quantity on the left is smaller than the quantity on the right. Although the distance 0P does not equal the sum $3 + 4$, it equals a sum involving these distances as we shall see below.

To get to the area associated with any point P on the plane we first lay off unit distances such as centimeters, starting from 0 and moving eastward along the $x$ axis. We label the endpoints of these distances 1, 2, 3, etc., just as we did in developing arithmetic. But here each number represents or equals the distance from 0 along the $x$ axis of the point it identifies. We now do exactly the same thing northward along the $y$ axis. We have thus divided both the positive $x$ axis and the positive $y$ axis into equally spaced intervals, each of unit length. At each point on the $x$ axis we construct a line parallel to the $y$ axis (perpendicular to the $x$ axis). We thus have an array of parallel north-pointing lines starting from the $x$ axis. We also construct an array of equally spaced parallel lines from the points 1, 2, 3, etc., on the $y$ axis pointing eastward. These two sets of lines cut each other at right angles and thus form an array of squares (like tiles on a floor) that cover the entire $x$-$y$ plane. We now define the area of a given section of the plane by the number of squares that cover this section.

We must be careful to assign the correct dimensions to the area, which is not just a number but a product of two numbers, each a distance. Taking the centimeter as the unit of distance we see that an area is centimeters times centimeters (cm × cm) or centimeters squared ($cm^2$). We now consider the north-south line passing through the point 1 on the $x$ axis and the east-west line passing through the point 1 on the $y$ axis; these two perpendicular lines intersect at the point $(1, 1)$ on the plane—the point whose X coordinate and Y coordinate are both 1. Together with the $x$ and $y$ axes these lines form a square, 1 cm on each side; therefore, the square has an area of 1 square centimeter. We call this quantity the unit area and measure any area by laying off this unit square as often as is necessary to cover completely the area being measured. This method, of course, requires dividing the unit area into halves, thirds, quarters, and so on, which is exactly what a floor man does when he tiles the floor of a house with tiles of a fixed size.

Considering now all the east-west parallel lines that pass through the integers on the $y$ axis and the north-south parallel lines that pass through the integers on the $x$ axis, we see that they form unit area squares that cover the entire plane; by counting these squares we obtain the area covered. Since area is a product of two perpendicular distances, we see that determining the area by counting these unit squares is really a multiplication operation, but it is multiplication on a higher level than the standard arithmetic multiplication. To illustrate these ideas numerically, we consider four parallel east–west lines passing through the points 1, 2, 3, 4 on the $y$ axis and three parallel north–south lines passing through the points 1, 2, 3 on the $x$ axis. Twelve unit squares are produced by these intersecting lines, which is the product of 3 and 4. We see then that the points on this plane constitute a multiplication table; if we label each point by the product of the number specifying the east–west line and the number specifying the north–south line passing through it, this product is just the area expressed in the unit areas enclosed by these two lines and by the $x$ axis and the $y$ axis.

This example shows us the intimate relationship between geometry and arithmetic; geometry becomes arithmetic and arithmetic becomes geometry. Whether we look at it one way or the other is merely a matter of personal preference. Geometry, however, widens our horizons in a way that arithmetic alone is unable to do. The reason for this difference is that geometry introduces dimensionality without which we cannot understand space at all. By introducing three sets of mutually perpendicular lines and labeling points on any three of these lines numerically (one line from each set) we impose an arithmetic on space. This feat represents one of the most remarkable intellectual achievements in the history of thought and was achieved by the French philosopher René Descartes.

Just as arithmetic leads to a deeper understanding and simplification of geometry, so geometry gives us answers to arithmetic problems without our having to do any arithmetic calculations. In particular, we consider the square root of numbers such as 2 which, themselves, are not the squares of rational numbers. To this end we draw that straight line through the origin 0, every point of which is at equal distances from the $y$ axis and the $x$ axis; this line divides (bisects) the

90° angle between the $y$ axis and the $x$ axis at 0 into two equal 45° angles. If, from any point on this angle bisector (the 45° line), we draw a north–south line parallel to the $y$ axis and an east–west line parallel to the $x$ axis, then these two lines and the $x$ and $y$ axes form a square, each side of which is the square root of the area of the square. As an example, we consider a point on the bisector which is at such a distance from 0 that the square formed by lines through this point, as described above, has an area of 7 cm$^2$. The distance of this point from either the $x$ or $y$ axis is $\sqrt{7}$ cm. In other words, we can get the square root of any number by constructing a square with any point on the 45° bisector at the upper right-hand corner of the square. If the measured area of this square equals the number involved, the distance of the point from the $x$ or $y$ axis is the square root of the number.

The area 2 is a special and particularly simple case because the point on the 45° line which is at the upper right-hand corner of the square (whose area is 2) is exactly 2 distance units along the 45° line, from the origin 0, so that the $\sqrt{2}$ is the distance, of this point in the given units of length, from the $x$ or $y$ axis. This can be deduced from the famous theorem of the great Greek philosopher Pythagoras, who developed his geometry and philosophy in the 5th century B.C.

Because of the great importance of the area concept in geometry, we have spent considerable time discussing it, noting that we can specify any area by a point on the plane, the product of whose $x$ and $y$ distances ($x$ and $y$ coordinates) equals the area of the square or rectangle whose upper right-hand corner coincides with the point that is chosen. If this point lies on the 45° line (the angle bisector) the figure is a square. Otherwise, the figure is a rectangle. In any case, the area is the product of the two sides that are at a right angle to each other. We now consider the area of a figure formed by four lines, two of which are parallel to each other and to the $x$ axis and the other two of which are parallel to each other but not to the $y$ axis. These two parallel lines cut the first two parallel lines at angles that are equal but smaller than 90°. The figure thus formed is a parallelogram. Its area is the product of the length of the longest side of the parallelogram and the perpendicular distance between these two lines. If we call the longest side of the parallelogram the base of the parallelogram and the perpendicular distance between these two long sides the altitude, then the area of a parallelogram is the product of its base and its altitude. We can easily demonstrate this point if we first consider the area of a triangle, which we now define.

To introduce the triangle we return to points on our $x$-$y$ plane and consider any point P defined or located by its coordinates $x$, $y$; that is, by its distance $x$ from the $y$ axis and its distance $y$ from the $x$ axis. As an example we take the point (3, 4) where $x = 3$ and $y = 4$ and draw a line from the origin 0 to this point P and another line from any other point S on the $x$ axis to the given point P. These two lines, which intersect at P and the line 0S, form a triangle which we designate as 0PS; the points 0, P, S where two lines of the triangle meet are called the vertices of the triangle. Most of the elementary plane geometry taught in our schools deals with the properties of triangles and theorems about triangle. Before we consider some of these basic theorems, however, we introduce a few definitions. We define a "right triangle" as one in which one of its three angles (formed by the three intersecting sides of the triangle) is a right angle (an angle of 90°). If one of the three angles is larger than 90°, the triangle is called "obtuse," and if all three angles are smaller than

90°, the triangle is called "acute." A triangle having three sides of equal length is called "equilateral" and if the triangle has two sides of equal length, it is called an "isosceles triangle." The triangle is called "scalene" if its three sides are of unequal length.

With these definitions, we can now prove the most important theorem about triangles. This proof requires the application of the Euclidean parallel axiom, which states that given a straight line and a point P outside this line, only one line parallel to the given line can be drawn through P. We now apply this axiom to our triangle 0PS (we always identify a triangle by its three vertices) in the $x$-$y$ plane, where the $x$ axis (the base 0S of our triangle) is the given line of the axiom and P is the point outside the line. We now draw through P a line (the only such line according to our axiom) parallel to the $x$ axis, that is, an east–west line which cuts the $y$ axis at a distance above 0 that exactly equals the distance of P above the $x$ axis. The two sides 0P and SP of the triangle cut this parallel line at P and form two angles with it. These two angles and the angle which the two lines form with each other at P (the P vertex of the triangle) add up to 180°, a straight angle, which therefore equals the sum of the three angles of the triangle. This is obvious if we note that we make half a complete turn (180° turn) if we stand at P and first look directly westward along the parallel line from P to the $y$ axis and then turn until we are looking along the line P0 (one of the sides of our triangle). In doing so we turn exactly through the angle formed by the side 0P and the parallel line described above. But this turn exactly equals one of the angles of the triangle, the angle which the side P0 of the triangle makes with the $x$ axis, which is parallel to the line through P. We now turn from the line P0 and look along the line PS, thus describing the angle of the triangle at the vertex P.

Finally, we turn from the line PS to the parallel line looking eastward through the angle which the side PS of the triangle makes with the parallel line, which is also the angle the line PS makes with the base 0S of our triangle, thus forming the third angle of our triangle. Thus, the total angle through which we have turned equals half a complete turn and hence 180°, but it also equals the sum of the three angles of the triangle. We have thus proved that the sum of the three angles of any triangle on our plane equals 180°.

This theorem is important because it seems to present a direct way of checking whether or not any particular space such as our three-dimensional space is Euclidean and obeys Euclid's axioms. We touched on this question in an earlier chapter but we consider it now more carefully. Our discussion of triangles above is highly idealized and our proof is an exercise in pure logic. Perfect planes, as infinitely thin, two-dimensional domains in three-dimensional space have no real existence nor do perfectly straight, one-dimensional lines and zero-dimensional points. Our plane triangles, consisting of "straight" lines and ideal "points" and lying on ideal flat "planes," therefore have no real existence. Proving the triangle theorem experimentally, therefore, by actual measurement of the angles is impossible, if not meaningless. But this theorem and the axiom that leads to it were very important in the evolution of geometry because an analysis of the axiom by 19th century mathematicians showed that it cannot be deduced from the other axioms and that without it Euclidean geometry is incomplete. Geometry advanced only at the expense of this axiom, which was replaced by two different axioms; these two axioms led to two different non-Euclidean geometries which we discuss in more detail later.

Two important concepts in the geometry of triangles are those of congruence and similarity. We say that two triangles are congruent if they can be superimposed on each other so that they cover each other exactly. This means that the two triangles must have certain minimal identical features or, put differently, each triangle is completely defined or specified by these features. Since the basic elements of a triangle are the lengths of its three sides and the magnitudes of the angles at its three vertices, different combinations of these quantities can be used to describe a triangle completely and therefore specify when two triangles are congruent (superimposable). One such combination is that of the three sides of a triangle; any triangle is completely defined or described if the lengths of its three sides are given. Expressed differently, two triangles are congruent if the lengths of their three sides are equal.

Instead of taking the lengths of the three sides of a triangle as the entities that determine a triangle, we can take the lengths of any two sides and the size of the angle formed by these two sides at the triangle vertex where they meet. Thus, any two triangles are congruent (can be exactly superimposed) if any two sides of one of the triangles are equal in length to the lengths of any two sides of the other triangle and the angles formed by these two sides in the two triangles are equal. A triangle is completely determined by two of its sides and the angle formed by these two sides. Finally, the length of any side of a triangle and the sizes of the two angles this chosen side of the triangle makes with the other two sides completely determine the triangle. Two triangles are congruent if two angles and the side included between these two angles in one of the triangles equal two angles of the other triangle and the side included.

These three statements listing the conditions for two triangles to be congruent are equivalent as can easily be demonstrated. Most of the problems in geometry that students meet in elementary courses in geometry can be solved with

the aid of any one of these three congruency theorems and
the concept of the similarity of triangles. Two triangles are
defined as similar if the three angles of one of the triangles
equal the three angles of the other triangle. As an example
of the application of the concept of similarity we note that
the shapes of all similar triangles are identical; they differ
only in size so that if all the sides of a triangle are increased
or reduced by exactly the same multiple we obtain a similar
triangle. If, in any triangle, we draw a line from a point on
one side of the triangle to a point on one of the other sides,
in such a way that the drawn line is parallel to the third side
of the triangle, we obtain two similar triangles—the original
large one and a smaller one embedded in the larger one.

We now introduce the area of a triangle which we do
most simply by considering the area of a square first. Returning to our $x$-$y$ plane we consider the point $x$, $y$ in the plane
with coordinates $(x, y)$; this point is $x$ units of distance from
the $y$ axis (east of the point 0). By drawing a line through
this point parallel to the $x$ axis and another line through it
parallel to the $y$ axis, we obtain the rectangle $x$ units wide
and $y$ units long. The area of this rectangle is $yx$ square units.
Taking a specific example, we consider the point in the $x$, $y$
plane which is 3 feet to the east of the $y$ axis and 4 feet to
the north of the $x$ axis so that our rectangle is 3 feet by 4
feet and its area is 12 square feet. We now draw the diagonal
from the origin 0 to the point $x$, $y$ (that is, to the point 3, 4
in our example). This line divides the rectangle into two equal
right triangles. Hence, the area of each of these triangles
equals exactly half the area of the rectangle. Alternatively
stated, the area of each triangle equals exactly half the product
$xy$ or, in our example, half the product $3 \times 4$. We now define
that side of the triangle which lies along the $x$ axis as the
base of the triangle and the side perpendicular to the $x$ axis

as the altitude of the triangle. In our specific example, the base is 3 feet and the altitude is 4 feet. We could, of course, have reversed these names and called the side lying along the $x$ axis the altitude and the other side the base, but this terminology is just a matter of personal preference. In any case, the area of our right triangle is one half its base times its altitude, or one half its altitude times its base.

Although we have derived this formula for a right triangle, it applies to triangles of any shape. We may then take any side of such a triangle as the base of the triangle and the altitude as the perpendicular distance from the vertex of the triangle whose base extends from the origin 0 to some point B at $b$ units of distance from 0. The length of the base is thus 0B measured along the $x$ axis. Let the distance $d$ along the $y$ axis from 0 to a point A on the $y$ axis be the altitude of the right triangle with vertices AB0. The line from the vertex B to the vertex A of the right triangle is called the hypotenuse of the right triangle. The area of this triangle is $(1/2)bd$, as already noted. Consider the line through the vertex A parallel to the $x$ axis and any point P, different from A, on this line. Every such point is at the same distance $d$ north of the $x$ axis. This distance $d$ is thus the altitude of any triangle formed by the lines from 0 to P and from B to P. Since such triangles also have the same base $0B = b$, they all have the same area.

We now deduce another formula for the area of a triangle using very general reasoning and elementary logic. This approach to mathematics is usually dismissed by mathematicians as lacking in rigor but it is extremely useful in that it leads one quickly to the desired result. This procedure can then be embellished, if required, with the fine print that a rigorous but often tedious proof demands. The area formula we have in mind contains the lengths of all three sides of the triangle, not just the base and altitude. If $a$, $b$, $c$ are the lengths of the three sides of a triangle, the length of its perimeter is $a + b + c$; the area of the triangle varies with the length of its perimeter: The larger the perimeter, the larger is the area. Since the perimeter is a length and not an area, it must be multiplied by a factor (another length) to give a length squared (which is an area). Of particular interest is that this factor must be equal to zero if the triangle reduces to a straight line (the area becomes 0) which occurs if the length of one of the sides of the triangle equals the sum of the lengths of the other two sides. Such a factor is $(b + c - a)$ which is zero for $a = b + c$. But this is true also of the factors $(a + b - c)$ and $(a + c - b)$. Since we want the area formula to be zero when $a = b + c$ or $b = a + c$ or $c = a + b$, we must use all three of these factors in the area formula. This procedure gives us the multiple product $(a + b + c)(a + b - c)$ $(a + c - b)(c + b - a)$, but this formula as it stands is incorrect because each factor is a length and the product is the fourth power of a length—length $^4$—whereas area is the square of a length. We must therefore take the square root of our multiple product to obtain an area. This procedure gives us the formula $\sqrt{(a + b + c)(a + b - c)(a + c - b)(c + b - a)}$. This formula has the basic features that the formula for the area of a triangle must have, but we still do not know whether or not the numerical coefficient 1 in front of the square root is correct. To check it we consider the area of a right triangle whose three sides have the lengths 3, 4, 5 units of length; the area of this triangle is 6 square units. In this triangle the perimeter $a + b + c = 12$ and $a + b - c = 3 + 4 - 5 = 2$, $a + c - b$ $= 3 + 5 - 4 = 4$, and $b + c - a = 4 + 5 - 3 = 6$ so that the square root above becomes $\sqrt{12 \times 2 \times 4 \times 6} = \sqrt{12 \times 12 \times 4}$ $= \sqrt{12^2 \times 4} = 24$. Since this quantity is four times larger than the actual area of our sample triangle we must take one-fourth of it to obtain the correct area 6. Our square root formula for the area of a triangle must therefore be multiplied by 1/4 to make it correct. Thus, the area of any triangle equals $(1/4)\sqrt{(a + b + c)(a + b - c)(a + c - b)(b + c - a)}$.

We complete our brief description of the geometry of triangles with a discussion and deduction of one of the most famous theorems in mathematics: the theorem for a right triangle. This theorem also deals with areas but not with the area of a triangle itself. We consider a right triangle in our $x$-$y$ plane, one side of which of length $a$ lies along the $x$ axis (the abscissa) eastward from the origin 0 to the point B. The other side of length $b$ lies along the $y$ axis (ordinate) northward from the origin 0 to a point A. The line of length $h$ from the point B on the abscissa to the point A on the ordinate is the hypotenuse of this right triangle. Suppose that this right triangle is the courtyard of a three-room house, with each room a perfect square; one room is built along the hypotenuse so that its floor space (area) is exactly $h^2$ units of area; the second room is built along the east side $a$ with floor space $a^2$ units of area; and the third room is built along the north side $b$ with floor space $b^2$ units of area. Pythagoras's theorem states that the floor space of the hypotenuse room equals the sum of the floor spaces of the other two rooms. We express this equality algebraically as the equation $h^2 = a^2 + b^2$. The square of the hypotenuse equals the sum of the squares of the other two sides.

We may deduce this theorem using very general arguments instead of presenting a rigorous geometrical proof; many such proofs are present in the mathematical literature. We begin by noting that the hypotenuse $h$ is always smaller than the sum of the other two sides $a$ and $b$, but the square $h^2$ of the hypotenuse is related to the sum of $a^2$ and $b^2$ in some way which we now analyze. Since $h^2$ is the square of a distance, it can only depend on the squares of distances that can be obtained from $a$ and $b$. Only three such quantities are available—$a^2$, $b^2$ and $ab$; therefore, $h^2$ can depend only on sums or differences of these three quantities. Since $h$ increases whether $a$ or $b$ increases, $h^2$ must depend on $a^2$ and $b^2$ in the same way. The algebraic formula for $h^2$ cannot contain a term involving $a^2$ that is different from the term involving $b^2$. Hence, as far as the dependence of $h^2$ on $a^2$ and $b^2$ alone goes, we must have $h^2 = pa^2 + pb^2$, where $p$ is a pure number which we must still determine. This formula for $h^2$ has the correct symmetry in its dependence on $a^2$ and $b^2$ but it does not contain the product $ab$. We shall show later that a formula containing $ab$ gives the wrong answer for $h^2$, but for the moment we keep it and consider the full equation $h^2 = pa^2 + pb^2 + qab$ where $q$ is a pure number. This is the only equation for $h^2$ that we can construct from $a^2$ and $b^2$ without violating our basic spatial axioms and the symmetry of the right triangle.

Since this equation is correct regardless of the size of $a$ and $b$, we apply it to a right triangle in which $b$ is extremely small (infinitesimal) compared to $a$ which is so if the vertex A of the triangle (the point on the ordinate; the $y$ axis) is brought infinitesimally close to the origin 0. The length of the hypotenuse $h$ then equals the length of the side $a$ of the triangle, as can be verified by inspection. But the two terms $pb^2$ and $qab$ on the right-hand side of the equation for $h^2$ are then zero since $b$ is zero (infinitesimal). For this triangle then the equation for $h^2$ becomes $h^2 = pa^2$ so that $h = \sqrt{pa}$ which means that $\sqrt{p} = 1$ (or $p$ itself equals 1) since $h$ must equal $a$, as explained above. Our equation for $h$ thus becomes $h^2 = a^2 + b^2 + qab$.

We must now analyze the role of the term $qab$. We consider two other triangles—one acute and one obtuse—in addition to the right triangle, in which the two sides $a$ and $b$ are of the same length as in the right triangle but with $b$ turned closer to $a$—closer to the $x$ axis (abscissa)—to form the acute triangle AB0 and with $b$ turned farther away from $a$ to form the obtuse triangle A0B. In the acute triangle, side $b$ points northeast and the angle it makes with side $a$ is smaller than 90°. In the obtuse triangle, side $b$ points northwest, and the angle it makes with side $a$ is larger than 90°. We note that as side $b$ of the triangle is brought closer to side $a$, side $h$ (the line from the point B on the $x$ axis to the point A which is to the right of the $y$ axis) gets smaller and smaller than the hypotenuse AB of the right triangle. This can only happen if the term $qab$ on the right-hand side of the equation for $h^2$ is negative, thus decreasing the expression $a^2 + b^2 + qab$ below its value $a^2 + b^2$. As side $b$ is turned closer and closer to $a$, $q$ becomes more and more negative.

Just the opposite happens to $q$ for obtuse triangles in which side $b$ is to the west of the $y$ axis; the line BA (side $h$ of the triangle) is longer than the hypotenuse of the right triangle, which means that the term $qab$ must be positive if side $b$ of the triangle is tilted westward away from the ordinate ($y$ axis). We thus learn something interesting about any triangle B0A; the coefficient $q$ of $ab$ in the equation for $h^2$ is negative (less than zero) for an acute triangle (the angle at the vertex 0 formed by the intersection of $a$ and $b$ is smaller than 90°), but as this angle increases, $q$ becomes less and less negative, approaching zero. When the triangle becomes obtuse (the angle at 0 exceeds 90°), $q$ becomes positive. This means that $q$ must be zero when the angle at 0 is 90°. Hence, for a right triangle $h^2 = a^2 + b^2$, the theorem of Pythagoras.

We have developed this theorem in some detail to illustrate the broader significance of the theorem than is indicated by its application to a right triangle. We come back to this theorem in the next chapter when we introduce the trigonometry of triangles. We complete this discussion with the note that the theorem of Pythagoras leads to another property of any point P on our $x$-$y$ plane. If the coordinates of this point are $x$ and $y$ (its distance from the $y$ axis is $x$ units and its distance from the $x$ axis is $y$ units), then the line $r$ drawn from the origin 0 to P (the distance of P from 0) is the hypotenuse of a right triangle whose two sides are $x$ and $y$. Hence, $r^2 = x^2 + y^2$ and $r = \sqrt{x^2 + y^2}$; the distance of any point on the plane equals the square root of the sum of its squared distances from the $x$ and $y$ axes.

# CHAPTER 6

# The Geometry of the Circle and Trigonometry

*Besides the mathematical arts there is no infallible knowledge, except it be borrowed from them.*
—ROBERT RECORDE

Of all the geometrical figures on the plane, the circle is the most important and interesting owing to its impact on mathematics, science, technology, and theology. From the earliest times, philosophers, metaphysicists, mystics, theologians, natural philosophers, and scientists saw the circle as an ideal figure which could be used to describe religious or natural phenomena. Thus, to the early theologians the circular shape of the celestial sphere marked the circle as the ideal, perfect figure to be associated with the abode of God and the angels. Nothing less than such perfection would suffice for the motions of the celestial bodies. To the ancient Greek philosophers and astronomers, the daily wheeling of the stars, the sun, the planets, and the moon across the sky in perfect circles emphasized the importance of the circle in celestial phenomena as did the circular disks of the sun and moon. This acceptance of the circle as the only admissible shape for the orbits of astronomical bodies led Ptolemy to his epicycle model of the solar system and Copernicus to the belief that Earth and other planets revolve around the sun in circles. Only after Kepler was forced to discard the circle in favor of the ellipse to obtain a correct orbit for Mars was the circle finally banished from astronomy as the only acceptable orbit.

Geometrical figures such as the circle, triangle, and line are important in mathematics and science because these forms tend to appear with great frequency in nature. Of even greater importance was the discovery by the Greeks that certain properties of these figures are both obvious and basic and that certain deductions can be made from the most obvious of these properties. "Surely if some new facts could be derived, these facts would apply to all those physical objects that possessed the basic properties in the first place. If the area of a circle could be shown by reasoning to be $\pi$ times the square of the radius, then the area of any circular piece of land should also be $\pi$ times the square of its radius. Further, perhaps by reasoning one could discover new facts which observation alone would not suggest. These advantages and many more the Greeks expected to derive from reasoning about common concepts on the basis of clearly evident facts. This is the germ of the great Greek discovery, perhaps the greatest discovery man has ever made. Reasoning can produce knowledge that not only covers a multitude of cases in one swoop but may produce physically meaningful information that is entirely unforeseen."[1]

The circle was also crucial to the successful application of mathematics to mechanical processes. The introduction of the wheel revolutionized technology and the use of circular motions of moving parts in machinery helped to produce the Industrial Revolution. Given the multitude of applications to which the wheel was ultimately put, it is not surprising that the invention of the wheel was made very early in human history: Inventions such as the wheel, the lever, and the wedge "were made on the basis of an instinctive and unreflecting knowledge of the processes of nature, and with the sole end of satisfaction of bodily needs. Primitive men had to build huts in order to protect themselves against the weather, and, for this purpose, had to lift and transport heavy weights, and so on. Later, by reflection on such inventions themselves, possibly for the purposes of instruction of the younger members of a tribe or the newly-joined members of a guild, these isolated inventions were classified according to some analogy. Thus we see the same elements occurring in the relation of a wheel to its axle and the relation of the arm of a lever to its fulcrum—the same weights at the same distance from the axle or fulcrum, as the case may be, exert the same power, and we can thus class both instruments together in virtue of an analogy. Here what we call 'scientific' classification begins."[2]

That the circle has an intrinsic abstract purity has not been lost upon those who have studied its properties. Indeed, it is the perfect figure because it is completely symmetrical. "In the one sense symmetric means something like well-proportioned, well-balanced, and symmetry denotes that sort of concordance of several parts by which they integrate into a whole. Beauty is bound up with symmetry."[3] The symmetry of the circle, like that of Euclid's line and plane, thus provided the yardstick against which the work of artisans and engineers alike could be compared. "Because of their complete rotational symmetry, the circle in the plane, the sphere in space were considered by the Pythagoreans the most perfect geometric figures, and Aristotle ascribed spherical shape to the celestial bodies because any other would detract from their heavenly perfection."[4]

Since the circle has been studied more thoroughly by mathematicians, going back to Pythagoras, Euclid, and Archimedes, than any other figure, it is fitting that we devote a separate chapter to it. We shall see that the study of the circle leads us to trigonometry more naturally than any other approach. Although we shall not emphasize motion in our study of the circle we note there that the circle describes one of the two extreme types of motion. These two types of motion may be described as follows: (1) motion in which no change in direction occurs, which is straight line motion such as that described by an automobile moving along a perfectly straight path; and (2) motion in which direction changes constantly by exactly the same amount for each unit of distance traveled but the speed remains constant, which is circular motion such as in an automobile in which the steering wheel is kept turned by exactly the same amount all the time and the speed is kept constant.

To study the geometry of the circle we go back to our $x$-$y$ plane and consider all points which are at exactly the same distance from the origin 0. The curve in the plane that connects all such points is called a circle and the origin 0 is the center of the circle. The distance of 0 from any point on a curve is the radius of the circle and the curve itself is called the circumference $c$ of the circle. Since we can choose any distance we please for the radius of our circle we use the letter $r$ to represent any desired value of the radius. The smaller the quantity $r$, the smaller is the circumference $c$ of the circle. If we choose a definite value for $r$ such as 5 inches, then every point on the circumference $c$ of this circle is exactly 5 inches from 0—the center of the circle. Here we emphasize the word "exact" because we cannot construct or draw a circle in practice since we cannot do anything with exact precision. The circle is thus an idealized figure that has no real existence. This is already indicated because the center 0 as a point is an idealized concept as is the circumference which, as an infinitely thin line (it has length but no width), is an ideal concept and cannot be constructed (drawn) in practice. The pointed leg of the most precise compass that may be used to construct a circle does not mark the center 0 of the circle as a point but produces a crater of finite size and the pen on the other leg of the compass traces out a line of finite width and not an infinitely thin curve. In spite of the nonexistence of a perfect circle we imagine that the figure we draw with our compass, with 0 at its center, is an ideal circle and we discuss this figure as such.

We note first that every point on the circumference of this circle is identified (located) on the plane by its $x$ and $y$ coordinates (its distances from the $x$ and $y$ axes). This means that the $x$ and $y$ coordinates of any such point must be related to each other and to the radius $r$ in such a way that the point lies on the circumference. This relationship is precisely given by the Pythagorean theorem. We see that this is so if we draw the radius $r$ from the center of the circle 0 to the point $x$, $y$ on the circumference. We then obtain a right triangle of which $r$ is the hypotenuse and the distances $x$, $y$ of the point from the $x$ and $y$ axes are the other two sides (previously called $a$ and $b$). The Pythagorean theorem then tells us that $r^2 = x^2 + y^2$. If $x = 3$, $y = 4$ are the coordinates of a point on the circumference of a circle we must have, for its radius, $r^2 = 3^2 + 4^2 = 9 + 16 = 25$ and $r = \sqrt{25} = 5$. In other words, the point $x = 3$, $y = 4$ lies on the circumference of a circle of radius 5. These numbers, of course, represent distances on some scale such as centimeters. Thus, a point whose distance from the $x$ axis is 4 cm and whose distance from the $y$ axis is 3 cm lies on the circumference of a circle whose radius is 5 cm. This relation is also true of the points ($x = 4$, $y = 3$),

$(x = -3, y = 4)$, $(x = -4, y = 3)$, $(x = 4, y = -3)$, $(x = 3, y = -4)$, $(x = -3, y = -4)$, $(x = 5, y = 0)$, $(x = -5, y = 0)$, $(x = 0, y = 5)$, $(x = 0, y = -5)$. Each set of parentheses stands for a single point. If $x$ is negative the point is to the left of the $y$ axis, and if $y$ is negative the point is below the $x$ axis. If both $x$ and $y$ are negative the point is to the left of the $x$ axis and below the $y$ axis. The points on the circumference are distributed symmetrically to the left and right of the $y$ axis and above and below the $x$ axis. The circumference is symmetrical with respect to its center; it has radial symmetry.

We now follow a tiny particle as it moves along the circumference of the circle from some starting point to some other point and deduce a relationship between the distance the particle moves along the circumference, and the angle the line (radius) from 0 to the first point makes with the radius from 0 to the end point. To be specific we consider the circle described above whose radius is 5 inches, and we start the particle from the point $(5, 0)$ on the $x$ axis. This particle is located 5 inches to the right of the $y$ axis and 0 inches away from the $x$ axis. If one end of a 5-inch string is attached to the origin 0 and the other end is attached to the particle, the string turns leftward toward the $y$ axis (ordinate) as the particle moves counterclockwise along the circumference; the angle the string makes with the $x$ axis thus increases as the distance the particle moves increases. The distance that the particle moves increases by exactly the same factor as the angle increases. If the string is rotated from the $x$ axis through 45°, one-eighth of a complete turn (360°), the end of the string moves along one-eighth of the complete circumference. If the string is rotated 90° (one-quarter of a complete turn), the end of the string moves along one-fourth of the circumference, and so on. We see that the length of the arc of the circumference between the point where it cuts the $x$ axis and the end of the string is proportional to (a multiple of) the angle. Since this length is a distance and angle is a pure number, the multiple (factor) that must multiply the angle to give us the arc length must be a length; it is the radius of the circle.

To see how the length of an arc of a circle is related to (depends on) the radius of the circle, we consider a long rod, one end of which is attached to the center of the circle so that the rod is free to rotate about the center which is the origin 0 of our horizontal $x$-$y$ plane (which we imagine to be covered with sand). If the rod has equally spaced vertical prongs on it which trace out arcs on the sand-covered $x$-$y$ plane, as the rod is rotated a given angle from the $x$ axis, the lengths of these arcs increase by exactly the same multiples as do the distances of the prongs as we move outward from prong to prong. If the first prong is a unit distance (e.g., 1 foot) from 0 and the second prong is 2 units from 0 and the third prong is 3 units, and so on, the length of the arc produced by the second prong is 2 times that produced by the first prong and the length of the third arc is 3 times that of the first arc, and so on. A complete discussion involving the angles and radii that define circular arcs reveals that the length of an arc is doubled if we double its angle or double its radius. If we double both the length and radius, we therefore increase the arc by a factor of 4. This means that the length of the arc depends on the product of the radius and the angle.

A simple way of expressing this idea by an elementary formula is to introduce a unit of angle which embodies the relationship between arc length and the product of angle and radius directly. We introduced the degree as the unit of angle by taking it as the 360th part of a complete turn, but we can just as well take any fraction of a complete turn as our unit.

But we simplify the formula for the length of the arc of any circle by replacing the degree by a larger unit which is somewhat smaller than one-sixth of a complete turn. The reason for this strange choice of the angle unit is that it greatly simplifies the geometry of the circle. This unit stems from a problem which arose from the analysis of the geometry of a circle; this problem greatly puzzled the ancient Greek philosophers, particularly Archimedes, who finally solved the problem.

To state the problem, we return to the string, one end of which is attached to the origin 0 of our $x$-$y$ plane so that the other end describes a circle as we rotate the string. We now consider an arc of this circle whose length, as measured from the $x$ axis along the circumference of the circle to the free end of the string, exactly equals the length of the string. The angle through which the string rotates as its free end moves from the $x$ axis to the end of this arc is the angle we use for one unit; it is called the radian.

The definition of the radian as our unit of angle is simple and straightforward; it thus presents no unnecessary complexities. The problem lies in determining the size of the radian (the fraction of a complete turn), which is equivalent to determining how much longer than its radius is the circumference of a circle if it is laid out as a straight line or how many times the radius of a circle can be laid off along its circumference.

Before considering the solution to this problem we now note that if the angle through which our string rotates from the $x$ axis is one radian, the arc described is exactly one radius in length. If the angle is 2 radians, the arc length is 2 radii, and so on. If the angle between the string and the $x$ axis is $\theta$ (the Greek letter "theta") radians, and $S$ is the length of arc between the $x$ axis and the free end of the string, then

$s = r\theta$, where $r$ is the radius of the circle (the length of the string).

We turn now to the solution of the problem stated above: How many times is the circumference of a circle larger than its radius or how many radians equal a complete turn? The early Greeks estimated this number to be 6 and so they equated the length of the circumference $c$ of a circle of radius $r$ to $6r$, which gives too small a circumference. Archimedes solved this problem in a very clever way which almost led him to the discovery of the calculus. Archimedes showed that the length of the circumference of a circle is larger than its radius $r$ by 2 times a certain number $\pi$ (the Greek letter "pi"). The circumference of a circle equals $2\pi$ times radius or $C = 2\pi r$. The number $\pi$, which we discussed briefly in the first chapter, is one of the most remarkable and important numerical constants that mathematicians and physicists meet in almost every phase of their work. Indeed, the frequency with which $\pi$ appears in the literature of any civilization is an excellent measure of its technological progress; it appears in the motions of the planets, in the propagation of light, in the flow of electricity, in the structure of atoms, and in chemistry, to mention but a few of its haunts.

Since $\pi$ is such an important number, mathematicians have spent many hours evaluating it, even though its precise numerical value does not need to be known to use it in the solution of various problems in mathematics and physics. As we stated in our first chapter, $\pi$ belongs to a group of numbers called transcendental numbers (they cannot be roots of algebraic equations with rational coefficients) which are members of a still larger set of numbers—the irrational numbers. The irrational numbers include such numbers as $\sqrt{2}$ (called algebraic numbers) as well as the transcendental numbers. The number $\pi$, which is somewhat larger than 3, cannot be expressed as a fraction (as a decimal with a finite number of decimal places); it is 3.14159..., where the dots represent an infinite number of integers which do not repeat themselves in any pattern. One hundred of these dots have been replaced by integers so that the value of $\pi$ is now known to about 100 decimal places. This simply means that $\pi$ can never be known exactly; it can only be approximated but the approximation can be made as close to the actual value of $\pi$ as we please since the actual value is given as a sum of an infinite number of terms, so that if we have time and patience enough we can add enough of these terms together to give $\pi$ to 100 or more decimal places.

The great 18th century German mathematician Gottfried Wilhelm Leibniz, credited with Newton as a discoverer of the calculus, expressed $\pi$ as the infinite series $\pi = 4 - 4/3 + 4/5 - 4/7 + \cdots = 4(1 - 1/3 + 1/5 - 1/7 \cdots)$, where the dots mean an infinite number of terms that are to be understood. Another such series is

$$\pi + 2 + \frac{1}{3} + \frac{3}{4 \times 5} + \frac{3 \times 5}{4 \times 6 \times 7} + \cdots$$

also

$$\pi = 3 + \left(\frac{1}{2}\right)^3 + \frac{9}{4 \times 5}\left(\frac{1}{2}\right)^5 + \cdots$$

Finding new ways of expressing $\pi$ as a series which can be used to compute $\pi$ to a large number of decimal places is a game which mathematicians play, but the value 3.14159... or 22/7, which is approximately 3.142, is accurate enough for all practical purposes.

Since the radius of a circle is contained in the circle's circumference exactly $2\pi$ times (approximately 6.28319 times), and the radian (angular measure) is the angle formed at the circle's center by the two radii that meet the circumference at the two ends of the arc whose length is one radius, then the total number of radians in $360°$ is also exactly $2\pi$. Thus, one radian equals $360°/2\pi$ or $57.2957°$. With this understood we see that if an object moves along the circumference of a circle a distance $s$ and the radius $r$ attached to it turns through the angle $\theta$ (expressed in radians, not degrees), then the distance $s$ equals the radius multiplied by the angle $\theta$ or $s = r\theta$. If the particle moves from the $x$ axis to the $y$ axis along the circumference, the radius turns through $90°$ and the arc described is $1/4$ of the circumference so that $90° = (1/4)2\pi = \pi/2$ and $180° = \pi$, and so on. We thus have the following general rule: If $A°$ is the magnitude of an angle in degrees, its magnitude in radians is $A°$ times $2\pi/360$ or $\theta$ (radians) $= A° \times 2\pi/360$. Thus, $45° = \pi/4$ radians, $30° = \pi/6$ radians, and so on. Note that if the radius of a circle is a unit of length (one centimeter, one inch, one foot, or any other unit), the length of any arc of the circumference of the circle exactly equals the angle in radians covered (subtended) by that arc at 0.

An important property of a circle is that the tangent (a line that touches one and only one point on the circumference of a circle) to any point on the circumference of a circle is exactly perpendicular to (makes a $90°$ angle with) the radius of the circle drawn to that point. Consider a circle whose center is at the origin 0 of our $x$-$y$ plane, and follow a particle (a point) as it moves along the circle's circumference. The particle's direction of motion changes continuously; the direction of the tangent to the point of the circumference occupied by the particle at any moment changes as indicated by the angle the tangent makes with the $y$ axis. This angle decreases from $90°$ to $0°$ as the particle moves along the circumference from the $x$ axis to the $y$ axis, and then increases again to $90°$.

We note further that if shadows of the point were cast on the $x$ and $y$ axes simultaneously, the two shadows would oscillate back and forth but they would be out of step. With the $y$ shadow at the origin 0, the $x$ shadow would be at its greatest distance (equal to the radius of the circle) from 0 and vice versa.

We may consider this idea from a different point of view. The position of any point on the circumference of a circle is given by its $x$ coordinate (its distance from the $y$ axis) and its $y$ coordinate (its distance from the $x$ axis). As the particle moves from point to point along the circumference (from the $x$ axis to the $y$ axis), the $x$ coordinate decreases steadily from $r$ (the radius) to 0 while the $y$ coordinate increases from 0 to $r$ and then the $x$ coordinate continues (becoming negative) to decrease while the $y$ coordinate begins to decrease. If we divide the $y$ coordinate of any point on the circumference by its $x$ coordinate, the quotient is a measure of the angle (equal to its complement) that the tangent to that point makes with the $x$ axis; it therefore specifies the direction of the tangent. We return to this point in the next chapter.

We now deduce the formula for the area of a circle, which is a special case of the area of a triangle. Since area is the square of a length, the area of a circle must be some multiple of the square of its radius $r$, which is the only length in a circle available to us. The formula for the area of a circle must equal the square of its radius multiplied by some numerical constant: area = constant $\times r^2$. To find this constant we consider first the area of one-quarter of the circle: the area of the quadrant defined by the arc of the circle from the $x$ axis to the $y$ axis. We may treat this as a kind of right isosceles triangle in which the two equal arms of the triangle equal the radius of the triangle and the base of the triangle is the arc just described. Since the two arms of the triangle (the $x$ and $y$ radii) meet at an angle of 90° or $\pi/2$ radians at 0, the length of the arc and, hence, the base of our triangle is $(\pi/2)r$. But the altitude of this triangle is just the radius of the circle itself which is the only line that can be drawn from 0 perpendicular to the arc. The area of this quarter sector of the circle is thus equal to (1/2)base × altitude $= (\pi/4)r^2$. Since the total area is 4 times this quantity, we obtain for the general formula for the area of a circle of radius $r$ the expression $\pi r^2$.

   We now see how the circle leads us to a different way of measuring an angle which uses lengths, even though the angle is a measure of rotation, not a length. To that end we consider a circle whose center is at the origin 0 and whose radius is a unit of length. Such a circle is called a unit circle. If we use centimeters as our unit of length, the radius of our circle is to be 1 centimeter. Consider now any point on the arc of this circle that extends from the positive part of the $x$ axis (the part to the right or east of 0) to the positive part of the $y$ axis (the part above or to the north of 0). This point is marked by its two coordinates $x, y$ (it is $x$ units of length to the east of 0 and $y$ units of length north of 0) where both $x$ and $y$ are smaller than one unit of length. Let the radius of this circle (one unit of length) drawn to the point $x, y$ on the circumference intersect the $x$ axis at 0 at an angle $\theta$. If we now drop a line from the point P parallel to the $y$ axis, this line cuts the $x$ axis at right angles and at a point whose distance from 0 (to the right of 0) exactly equals $x$. The length of this line from the $x$ axis to the point P exactly equals $y$. As the point P moves along the circumference of the circle from the $x$ axis to the $y$ axis, three things, which are related to each other, happen. First, the angle $\theta$ which the unit radius from origin 0 to P makes with the $x$ axis increases from 0° to 90°. Second, the distance $x$ of P from the $y$ axis (the $x$

coordinate of P) decreases from 1 to 0. Third, the distance $y$ of P from the $x$ axis (the $y$ coordinate of P) increases from 0 to 1. This means that we can use any one of three different distances to represent or measure the angle: (1) Since the length $s$ of the arc along the circumference of the circle from the $x$ axis to P equals the radius of the circle times the angle $\theta$ and the radius $r$ is 1, we have $s = r\theta = \theta$, so that the angle $\theta$ in radians equals the length of $s$. (2) Since the distance $x$ (the $x$ coordinate of P) decreases steadily from 1 to 0 as $\theta$ increases from 0° to 90°, so that each value of $x$ corresponds to a definite value of $\theta$, we can use the distance $x$ to represent the angle $\theta$. We call this distance the cosine of the angle $\theta$ which is written as cos $\theta$. Note that neither the angle nor the cosine of the angle is a distance; they are pure numbers (they have no spatial dimensions) that (in the unit circle) numerically equal the length of the arc and the $x$ distance of P respectively as described above. (3) Just as we can use the $x$ coordinate of P to identify or measure the angle $\theta$, we can use the $y$ coordinate of P (its distance from the $x$ axis) to do the same thing. As the angle $\theta$ increases from 0 to 90° ($\pi/2$ radians), the $y$ coordinate increases from 0 to 1 so that for each value of $\theta$, the distance $y$ has one and only one value. We call this distance the sine of the angle $\theta$, which we write as sin $\theta$.

All that we call trigonometry is contained in the definition of the sine or the cosine of an angle. Indeed, we may say that either of these definitions is all there is to trigonometry. All the theorems that are associated with trigonometry can be deduced from either of these definitions. Before we illustrate this point by deducing a few elementary trigonometric theorems (sometimes called laws) we introduce the sign ("+" or "−") convention that mathematicians use in dealing with angles and the signs we must attach to the cosine and sine of an angle for negative angles or angles larger than 90°. We recall that in our number system we assign the "+" sign to numbers that label points on our number line that lie to the right of the origin 0 and the "−" sign to points that lie to the left of 0. These signs also acquired an operational significance in arithmetic in that the "+" sign requires that we add steps together if we move to the right and subtract steps if we move to the left. We do the same thing with angles, recalling that an angle is a measure of rotation. We assign a "+" or "−" sign to an angle depending on whether the angle represents a counterclockwise or a clockwise rotation. If the point P on the circumference of our unit circle moves from the positive $x$ axis to the positive $y$ axis, the unit radius attached to it turns counterclockwise and the angles through which the radius turns are called positive; if the point P moves along the circumference clockwise from the positive $x$ axis to the negative $y$ axis, the angles through which the unit radius turns are called negative.

The $x$ and $y$ axes divide the $x$-$y$, plane into four quadrants. In the first quadrant, the angle ($\theta$) that the unit radius describes as it rotates counterclockwise from the $x$ to the $y$ axis, increases from 0° to 90°. In the second quadrant, the unit radius turns from the positive $y$ axis to the negative $x$ axis and the angle $\theta$ increases from 90° to 180°. In the third quadrant, the angle $\vartheta$ increases from 180° to 270°, and in the fourth quadrant, the angle $\theta$ increases from 270° to 360°. If instead of rotating the unit radius counterclockwise, we rotate it clockwise, starting from the $x$ axis, the angles described by this radius are all negative. Thus, the angle −90° in the fourth quadrant corresponds to the positive angle 270°, the angle −180° corresponds to the positive angle 180°, the angle −270° corresponds to 90°, and the angle −360° corresponds to the angle 0. Finally, we note that the sine and cosine of $\theta$ can

be positive or negative depending on the quadrant in which
$\theta$ lies. Both the sine and cosine of $\theta$ in the first quadrant are
positive since $x$ (which gives the cosine) and $y$ (which gives
the sine) are positive. In the second quadrant (between 90°
and 180°) the cosine becomes negative (it is 0 for 90°) because
for a point on the circumference of the unit circle in the
second quadrant the $x$ coordinate is to the left of 0 and,
therefore, it and the cosine are negative. The sine is still
positive in the second quadrant because the $y$ coordinate of
the point (which gives the sine) is above the $x$ axis and
therefore positive. In the third quadrant both the sine and
cosine are negative. Finally, in the fourth quadrant the sine
is negative and the cosine is positive.

Having introduced the basic elements of trigonometry
(definitions of the sine and cosine of an angle) we can now
deduce some useful trigonometric theorems. We return to the
point P on the circumference of the unit circle whose coordinates (distances from the $x$ and $y$ axes) are $x$ and $y$, respectively. The unit radius from the origin 0 to the point P divides
the right angle formed by the $x$ and $y$ axes at 0 into the angle
$\theta$ which the radius makes with 0 and the angle $(90° - \theta)$
which the radius makes with the $y$ axis; this angle is called
the complement of the angle $\theta$ and its cosine is measured by
how far P lies above the $x$ axis—the $y$ coordinate of P, which
is also exactly the definition of the sine of $\theta$. Thus, the cosine
of $(90° - \theta)$ equals the sine of $\theta$ so that we have $\cos(90° - \theta) = \sin\theta$. In the same way we see that $\sin(90° - \theta)$, given
by the $x$ coordinate of P, equals $\cos\theta$ so that if we know the
sine and cosine of an angle, we know the cosine and sine of
its complement.

But knowing both the sine and cosine is not necessary
to discover all we want to learn about an angle; all we need
to know is the sine alone (or the cosine alone) to deduce all

the properties of angles. We consider the right triangle formed by the unit radius from 0 to the point P on the circumference of the unit circle, by the line parallel to the $y$ axis from P to the $x$ axis, and by the $x$ axis between 0 and the point on the $x$ axis where the line from P meets it. We thus have a right triangle whose hypotenuse is 1 unit of length (the radius of the unit circle) and whose other two sides (perpendicular to each other) are the coordinate distances $x$ and $y$ of P. The theorem of Pythagoras now tells us that $1^2 = x^2 + y^2$ (the square of the hypotenuse equals the sum of the squares of the other two sides). But $x^2 = \cos^2 \theta$ and $y^2 = \sin^2 \theta$ so that we have $\cos^2 \theta + \sin^2 \theta = 1$ or $\sin^2 \theta = 1 - \cos^2 \theta$. Hence, $\sin \theta = \sqrt{1 - \cos^2 \theta}$ and $\cos \theta = \sqrt{1 - \sin^2 \theta}$. If we know the sine of an angle, we can thus calculate its cosine and vice versa.

We now introduce another line associated with the unit circle, whose length can also be used to identify or measure the angle $\theta$. We again consider the point P on the circumference of the unit circle in the first quadrant and draw the line from 0 right through P and extending beyond it. At the point on the $x$ axis where the arc of the circumference from P to the $x$ axis cuts the $x$ axis we construct the line which is exactly perpendicular to the $x$ axis (exactly parallel to the $y$ axis). This is the tangent of the unit circle at that point. Consider now the length of this tangent from the $x$ axis to the point where it intersects the line from 0 through P. This length is called the tangent of the angle $\theta$ (written as tan $\theta$). As the point P gets closer and closer to the $y$ axis (the angle $\theta$ increases), the point on the tangent line where the line from 0 through P cuts the tangent line moves farther and farther away from the $x$ axis. In other words, tan $\theta$ increases from 0 to ever larger values as $\theta$ increases from 0 to 90°, becoming infinite for $\theta = 90°$. This definition of the tangent of $\theta$ (tan $\theta$) is equivalent to the quotient $y/x$ (the $y$ coordinate of P divided by its $x$ coordinate), from which we deduce the basic relationship $\tan \theta = \sin \theta / \cos \theta$ since $y$ is the sine of $\theta$ and $x$ is its cosine.

We have introduced the sine, cosine, and tangent of an angle by considering a point on the circumference of a unit circle but these definitions and the various relationships apply to all angles whether they are associated with the unit circle or not. If we work with a circle whose radius $r$ is larger or smaller than our unit of length, we choose a point P on its circumference as we do for the unit circle and measure its $x$ and $y$ coordinates (its distances from the $x$ and $y$ axes). These distances are not the cosine and sine of the angle $\theta$ that $r$, drawn from 0 to P, makes with the $x$ axis. But we obtain the cosine and sine of $\theta$ by dividing the $x$ and $y$ coordinates of P by $r$. Thus, for circles with radii not of unit length, $\cos \theta = x/r$ and $\sin \theta = y/r$ and we must divide by $r$ the length of the tangent line described above to obtain $\tan \theta$. This shows that the sine and cosine are dimensionless, because they are distances divided by distances.

We now consider any triangle whatsoever whose three pairs of sides intersect, at the vertices A, B, C, forming the three angles, which we label A, B, C, the same as the vertices, each of which can be identified by its sine or cosine. Consider the angle A formed by the intersection of the two sides AB and AC of the triangle. We now lay off along AB the unit length from A to some point P (the distance AP is a unit of length) and from the point P we drop a line perpendicular to the side AC. If D is the point on the side AC where the perpendicular line from P hits the line AC, then the distance AD, measured in our chosen units of length, is the cosine of the angle at A and the length PD (expressed in the same units) is the sine of the angle A. The quotient PD/AD is the tangent of A. We can do this with any angle whether it belongs to a triangle or not. In any right triangle the sine of either of the two angles formed by the hypotenuse and the other two sides is the length of the side opposite the angle divided by the hypotenuse and the tangent of that angle is the length of the side opposite that angle divided by the length of the adjacent side.

Two theorems are used in the trigonometry of triangles and in solving the triangle problems that authors of books on trigonometry delight in placing throughout their books. One of these theorems is called the law of sines and the other is called the law of cosines. The law of sines states that in a triangle the length of a side divided by the sine of the angle opposite that side is the same for all three sides and the angles opposite them. Mathematicians state this rule by saying that the length of any side of a triangle is proportional to the sine of the angle opposite that side. In applying this theorem to a triangle, one must express the lengths of the three sides in the same units of length. The sines of the angles are dimensionless (pure numbers) so we do not have to worry about them.

If the lengths of the three sides of a triangle are $a$, $b$, $c$ and they are opposite the angles (vertices) A, B, C respectively, then the law of sines can be written algebraically as $(\sin A)/a = (\sin B)/b = (\sin C)/c$ or $\sin A/\sin B = a/b$; $\sin A/\sin C = a/c$; $\sin B/\sin C = b/c$. Before this law can be used numerically, the numerical sines of angles between $0°$ and $90°$ must be known. This need led mathematicians to prepare numerical tables of the sines of angles in steps of one-thousandth of a degree (or smaller) for angles between $0°$ and $90°$. Later we shall see just how such tables are calculated numerically and how sines of angles that lie between two steps can be calculated from the tables by an arithmetic process called extrapolation. Without going into any details about these tables at this point, we note how the sines and cosines of certain angles can be calculated with elementary geometry and a bit of arithmetic. We return to our unit circle in the $x$-$y$ plane with its center at the center 0 and consider the unit radius from 0 which cuts the 90° angle at the origin 0 into two equal angles (45° each). This angle bisector cuts the circumference of the circle at a point P which is midway between the $x$ and the $y$ axes so that the $x$ and $y$ coordinates of this point are equal. This means that the sine and cosine of the 45° angle that the unit radius makes with the $x$ axis are equal. Since the square of the cosine plus the square of the sine of any angle equals 1, we have $\cos^2 45 + \sin^2 45 = 1$ so that $2 \sin^2 45 = 1$ or $\sin^2 45 = 1/2$. Thus, $\sin 45 = \cos 45 = \sqrt{2}/2$ so the sine and cosine of 45° are both approximately 0.707.

As another example we consider the right triangle formed by the unit radius (the hypotenuse of the triangle) drawn to a point on the circumference which is one-twelfth of the way around the circumference starting from the $x$ axis and the $x$ and $y$ coordinates of this point as the other two sides. This right triangle is a 30°, 60°, 90° triangle since the unit radius makes an angle of 30° with the $x$ axis and an angle of 60° with the $y$ axis. From the geometry of the equilateral triangle one can show that the side opposite the 30° angle (the $y$ coordinate) is one-half the radius ($y = 1/2$) which means that $\sin 30° = 1/2$. To obtain the cosine of 30°, which equals the $x$ coordinate of the point on the circumference, we apply the theorem of Pythagoras to our right triangle, noting that its hypotenuse is 1 (the unit radius) and the $y$ coordinate is 1/2 so that $x^2 + y^2 = 1 = x^2 + 1/4$ or $x^2 = \sqrt{3}/2$. Hence, $\cos 30° = \sqrt{3}/2$. Since 60° and 30° are complementary angles (their sum is 90°), the cosine of 30° equals the sine of 60° and vice versa.

From our knowledge of the sines and cosines of 30°, 45°, and 60°, we can deduce the sines and cosines of other angles by using two other trigonometric formulas which we state without proof. These are the formulas for the sum (or difference) of two angles; thus, if $\theta$ and $\phi$ are two angles (they may be the same or different) we have $\sin (\theta + \phi) = \sin \theta \cos \phi + \cos \theta \sin \phi$ and $\sin (\theta - \phi) = \sin \theta \cos \phi - \cos \theta \sin \phi$. Moreover, $\cos (\theta + \phi) = \cos \theta \cos \phi - \sin \theta \sin \phi$ and $\cos (\theta - \phi) = \cos \theta \cos \phi + \sin \theta \sin \phi$. That these formulas are very useful can be seen when we first consider the sine and cosine of 2 times an angle. Thus, $\sin (\theta + \theta) = \sin 2\theta = \sin \theta \cos \theta + \cos \theta \sin \theta = 2 \sin \theta \cos \theta$. The sine of twice an angle is twice the sine times the cosine of that angle. For the cosine of 2 times an angle we have $\cos (\theta + \theta) = \cos 2\theta = \cos \theta \cos \theta - \sin \theta \sin \theta = \cos^2\theta - \sin^2\theta$.

The usefulness of these formulas is illustrated by a few numerical examples. Thus,

$$\sin (75°) = \sin (45 + 30) = \sin 45 \cos 30 + \cos 45 \sin 30$$
$$= (\sqrt{2}/2)(\sqrt{3}/2) + (\sqrt{2}/2)(1/2)$$
$$= (\sqrt{2}/4)(\sqrt{3} + 1) = (1.414/4)(1.732 + 1)$$
$$= 0.353(2.732) = 0.964$$

We note also that

$$\sin 15° = \sin (45 - 30) = \sin 45 \cos 30 - \sin 30 \cos 45$$
$$= (\sqrt{2}/2)(\sqrt{3}/2) - (1/2)(\sqrt{2}/2)$$
$$= \sqrt{2}/4(\sqrt{3} - 1)$$

which is the same as $\cos 75°$, which it should be because 15° and 75° are complementary angles.

Later we show how the sine and cosine of any angle can be computed to any degree of accuracy we please (to any approximation) by using certain infinite series (sums of an infinite number of terms) that mathematicians have discovered, but here we return to the law of sines of the angles of a triangle to show its usefulness in surveying. The sine law, we recall, states that the sine of any angle of a triangle divided by the length of the side of the triangle opposite that angle equals the quotient of the sine of either one of the other two angles divided by the length of the side opposite that angle. If we know the length of any side of a triangle and the angles on the ends of this length, we can compute the lengths of the other two sides. To see how this is applied, we consider a tree which is located on the other side of a river from the viewpoint of a surveyor who wants to determine the distance of the tree. He first measures (paces off) a distance (such as 30 feet) along a straight line on his side of the river to some convenient point. He then uses his surveyor's transit (a special telescope) to sight the tree from his new position and measures the angle between this direction to the tree and the direction of the line he paced off by rotating the transit from the direction to the tree to the direction of the line back to his initial position. Let this angle be 45° and let the angle between the paced-off line and the direction of the tree from his starting position be 60°. This means that the two directions to the tree meet at an angle of 75°. The surveyor thus has a 45°, 60°, 75° triangle with the side opposite the 75° angle equal to 30 feet. Since his initial distance from the tree (which he is seeking) is opposite the 45° angle, he obtains the following equation using the law of sines:

distance sought/sin 45° = 30 feet/sin 75° or distance sought

$$= (\sin 45°/\sin 75°)(30 \text{ feet})$$
$$= \sqrt{2}/2 \times 30/0.964 \text{ feet}$$
$$= (0.707/0.964)(30) = 22 \text{ feet}$$

This procedure, called triangulation, is used extensively in surveying large areas on the Earth's surface. The surveyor must use a series of small connected triangles to cover the areas because the Earth's surface is a sphere on which a large triangle consisting of straight lines cannot be drawn.

We now turn to the law of cosines for triangles. This law is really the theorem of Pythagoras extended to any triangle. To see how this is to be done we return to our unit circle with its center at 0 (the origin of our coordinate system) and consider the triangle formed by the unit radius from 0 to a point P on the circumference in the first quadrant, the line from P to the point 1 on the $x$ axis (where the circumference cuts the $x$ axis), and the radius of the circle along the $x$ axis from 0 to point 1. The vertices of this triangle are O, P, 1, and two of its sides are radii of length 1 (an isosceles triangle). The third side is the chord of the circle from the point P on the circumference to the point 1 on its circumference; the length of this side (the chord) increases from 0 to $\sqrt{2}$ as the angle $\theta$ that the unit radius from 0 to P makes with the $x$ axis increases from 0° to 90°. When the angle is 90°, we have a right triangle, the hypotenuse of which is the chord from P (on the $y$ axis) to the point 1 on the $x$ axis. The theorem of Pythagoras now tells us that $(\text{chord})^2 = (01)^2 + (0P)^2$. But each of the two sides 01 and 0P is 1 so that we have $(P1)^2 = 1^2 + 1^2 = 2$ and the chord $(P1) = \sqrt{2}$. If the point P is anywhere on the arc of the unit circle between the $x$ and $y$ axes, the triangle is not a right triangle. The chord P1 is smaller than $\sqrt{2}$ so that we must have $\text{chord}^2 = 1^2 + 1^2 - q = 2 - q$ where $q$ (the square of a length) must decrease from 2 to give a value 0 for the chord when P is on the $x$ axis and coincident with point 1, and the value $\sqrt{2}$ when P is on the $y$ axis. The only quantity that has the dimensions of the square of a length and changes its magnitude the way $q$ must to give the correct length of the chord P1 is the product $(2 \cos \theta)(\text{side } 01)(\text{side } 0P)$.

We thus have the general theorem of Pythagoras $(\text{chord})^2 = (01)^2 + (0P)^2 - (2 \cos \theta)(01)(0P)$. Since in our present example P is a point on the circumference of the unit circle, the chord is P1, and the other two sides of our triangle are unit radii (both 0P and 01 are 1), we have $(P1)^2 = 1^2 + 1^2 - (2 \cos \theta) \times 1 \times 1 = 2 - 2 \cos \theta$, so that $(P1)^2$ varies from 0 for $\theta = 0°$, when $\cos \theta = 1$, to 2 for $\theta = 90°$ when $\cos \theta = 0$. When the point P is in the second quadrant to the left of the $y$ axis, the length of the chord P1 increases from $\sqrt{2}$ when $\theta = 90°$ to 2 when $\theta = 180°$ and the distance from P to point 1 equals two radii. This is exactly stated by the formula for $(P1)^2$ because the cosine of the angle $\theta$ is negative in the second quadrant (the $x$ coordinate of P is to the left of the $y$ axis and thus negative) and the third term in the equation for $(P1)^2$ becomes positive. For $\theta = 180°$ whose cosine is negative, we have $(P1)^2 = 1 + 1 - 2(\cos 180°)(1)(1) = 2 + 2 = 4$ and $P1 = \sqrt{4} = 2$.

We have presented the cosine law here for the very special case of a triangle having two sides equal to 1, but it is valid for any triangle. We state it now for a triangle whose three angles are A, B, C and whose sides opposite these angles have the lengths $a$, $b$, $c$ respectively: $a^2 = b^2 + c^2 - 2(bc) \cos A$, or $b^2 = a^2 + c^2 - 2(ac) \cos B$, or $c^2 = a^2 + b^2 - 2(ab) \cos C$. Thus, the length of any side of a triangle equals the sum of the squares of the lengths of the other two sides minus twice the cosine of the angle opposite the first side times the product of the lengths of the other two sides.

We have spent considerable time discussing the trigonometry of triangles and developed the sine and cosine concepts as they apply to triangles, but these concepts extend far beyond the triangle and appear extensively in all branches of science and technology, particularly in the mathematical descriptions of phenomena involving periodic motion such as that of the bob of a pendulum oscillating back and forth or that of a planet moving around the sun or that of the vibrations of a string. But the sine and cosine are used extensively in the analysis of any kind of motion as we can understand if we note that the motion of a particle in a plane can be described completely in our $x$, $y$ coordinate system by following its $x$ and $y$ coordinates separately. If the particle at any moment is at a point P whose distance from the origin 0 is $r$ and if the line $r$ from 0 to P makes the angle $\theta$ with the $x$ axis, then the $x$ coordinate of the particle is $r \cos \theta$, its $y$ coordinate is $r \sin \theta$, and the direction of the particle's motion is given by $\tan \phi$ where $\phi$ is the angle the tangent at P makes with the $x$ axis (the slope of its path). Breaking up the motion of a particle in this way greatly simplifies the solution of a very important practical problem: the launching of satellites.

The sine and cosine also enable us to analyze mathematically the properties of waves. To see what is involved in such an analysis we first consider two phenomena, one simple and the other complex. However, both are governed by the same basic physical principles. The simple example involves dropping a stone in still water on which a cork is floating while the second concerns an electron in a wire along which a radio wave is passing. Observing the first example, we note that when the stone (which is assumed to be perfectly round) strikes the water, vertically oscillating concentric circles of water of ever-increasing radii move outward from the spot where the stone entered so that one point on the circumference of the first (largest) circle eventually meets the cork. When this occurs, the cork begins to oscillate up and down and continues to do so until the last circle of water has passed it.

The concentric circles are called surface water waves. We note that in this motion the cork does not move horizontally at all; it only moves up and down like the water itself.

We consider now the electron in the wire; it responds to the oscillating electric and magnetic fields which constitute the radio wave (an electromagnetic wave) that passes it. The electron also oscillates in step with the electromagnetic wave and a pulsing electric current is generated which moves through the wire into a receiver such as a radio.

Before we see how the sine and cosine are used to describe this wave phenomenon mathematically, we note that the wave transmits energy from the source of the wave to the receiver. In the example of the cork, the energy comes from the falling stone; in the case of the electron, the energy comes from the oscillating electrons in the radio station which emit radio waves. To see how the sine and cosine describe the water wave (the vertical oscillations that are transmitted horizontally), we return to our unit circle and focus on the shadows on the $x$ and $y$ axes of a particle moving with constant speed around the circumference of a circle. The angle $\theta$, which the unit radius from 0 to the particle makes with the $x$ axis, increases steadily from 0° to 360° as the particle revolves around 0 once; this is repeated for each trip around the circle. The shadow of the particle on the $y$ axis oscillates up and down (north and south) relative to the $x$ axis. Since the distance of this shadow from 0 along the $y$ axis is the sine of $\theta$, we may describe it (the distance) at any moment $t$ (here, $t$ stands for the time in seconds that has elapsed from the moment the particle started revolving from the $x$ axis) as sin $2\pi n t$. Here $n$ is the number of times the particle revolves in a second and 360° is replaced by $2\pi$ radians. The quantity $2\pi n t$ is just another way of writing $\theta$ at the moment $t$ if the particle revolves $n$ times per second because $\theta$ then increases by 360° or $2\pi$ radians for each $1/n$th of a second. We may, of course, watch the shadow of the particle on the $x$ axis, which is its $x$ coordinate and therefore given by the cosine of $\theta$. Thus, we can describe the oscillations of the $x$ shadow by the expression $\cos(2\pi nt)$.

These oscillations do not represent or describe a wave because they are oscillations with respect to only one point on the $x$ axis—the origin 0. But we obtain a wave moving to the right at a speed $v$ if we move 0 to the right at a speed $v$ and replace $t$ by $t - x/v$, where $x$ is the distance the origin 0 moves in the time $t$. Thus, $\sin 2\pi n(t - x/v)$ describes a wave moving to the right at a speed $v$ and oscillating vertically $n$ times every second. This is just how a surface water wave behaves. Note how the sine (as well as the cosine) sheds its elementary simple significance—which stems from its lowly birth as a mathematical entity to describe an angle—and becomes a very powerful mathematical tool in physics. It evolves from a rag-clad scrubwoman to an elegant princess. Since electromagnetic waves (X rays, light, infrared rays, microwaves, radio waves, etc.) consist of electric and magnetic fields oscillating at the same rate at right angles to each other, we can represent two such oscillations together by using the sum of the sine and the cosine (the simultaneous oscillations of the two shadows). If we now move our $x, y$ coordinate system out of the plane of the paper so that the origin 0 (the center of our unit circle) moves along a line perpendicular to the plane at the speed of light, the sum of the oscillations of the two shadows (described by the sine and cosine) is the mathematical representation of an electromagnetic wave.

The waves described above are pure waves in the sense that they consist of waves vibrating or oscillating at a single rate or frequency, which is given by the number $n$. In general, however, vibrations or waves consist of a mixture of vibrations, each with its own $n$ value. Examples from various natural phenomena illustrate this point very nicely. Consider first a tuning fork which is designed to emit a note of a single pitch when struck. The sound is then a pure acoustical wave whose frequency (the number $n$) has a very definite value; the larger this number $n$, the greater the frequency of the wave (the more rapid are its vibrations) and the higher the pitch of the sound. An actual sound produced by a combination of musical instruments or in any other way consists of mixtures of vibrations (called harmonics); special devices can be used to break up this sound, however complex, into a sum of pure vibrations with their individual frequencies ($n$ values). This process, called harmonic analysis, is equivalent mathematically to expressing the acoustical disturbance as a sum of sines and cosines with different $n$ values and different coefficients in front of the sine and cosine terms to express the different intensities (loudness) of the different acoustical waves (harmonics) that combine to give the sound produced. The quality of musical sounds depends on how such harmonics are combined to produce the music.

Ordinary (visible) light consists of a mixture of many different electromagnetic vibrations, ranging from 1000 trillion vibrations per second (violet light) to 500 trillion vibrations per second (red light); the propagation of a beam of ordinary light (white light) is therefore described mathematically by a sum of sines and cosines with different $n$ values and different coefficients. A prism through which an ordinary beam of light passes spreads the light out into its pure colors (Newton's famous experiment), each of which is propagated as a train of monochromatic waves (a sine–cosine combination) with a definite numerical value for $n$. A laser beam is a train of pure monochromatic electromagnetic waves described by the sum of a single sine and cosine with a single value for $n$ (frequency).

Although we have described a wave as the propagation of a vibration oscillating with a definite frequency, we can use the wavelength of the wave (the distance between two successive crests of the wave) instead of the frequency because wavelength and frequency are related to each other by a simple equation. If the wavelength of a light wave is $\lambda$ (the Greek letter "lambda"), $n$ is its frequency (number of oscillations per second), and $c$ is the speed of the wave, then $(\lambda)(n) = c$; the product of the wavelength and the frequency of the wave always equals its speed, which is the same for all colors (wavelengths) in a vacuum.

We have seen that the usefulness and importance of the sine and cosine in describing the propagation of waves transcend their original mathematical function of enabling us to express the magnitudes of angles in terms of the ratios (quotients) of distances. But the discovery by the great French mathematician and physicist Jean-Baptiste Fourier, in the early 19th century, that any algebraic expression can be represented as an infinite sum of sines and cosines (a Fourier or trigonometric series) underscores these trigonometric functions, as the sine and cosine are called, and gives them a preeminence enjoyed by no other mathematical functions. Thus, Fourier series are very useful in analyzing the conduction of heat in a solid, in radioengineering, and in the conduction of electricity in complex circuits, to name but a few areas in which trigonometric series enable us to solve very complex problems.

Mathematicians, recognizing the importance of the trigonometric functions (sine, cosine, and tangent) for mathematics and science very early in the history of these subjects, developed mathematical procedures for calculating the values of the cosine and sine of any angle to any desired accuracy (to any number of decimal places). Using these procedures, mathematicians have constructed extensive numerical tables in which the values of the sines and cosines are given for angles from 0° to 90° in very small steps (tenths of a degree). The mathematical device that is used to calculate the sine and cosine of any angle is to express each of these trigonometric functions as an infinite series (an infinite sum), each term of which is the angle raised to a different power. To indicate the mathematical nature of the sine and cosine series, we return to our unit circle that has served us so well and again consider a point P on its circumference in the first quadrant. If the radius from the origin 0 to P makes an angle $\theta$ with the $x$ axis, the distance of P from the $x$ axis equals sin $\theta$. If P is very close to the $x$ axis, this distance (from P to the $x$ axis) very nearly equals the length of the arc of the circle from the $x$ axis to P. But this arc length measured in the same units of length in which the radius of the unit circle is 1, equals the size of the angle $\theta$, expressed in radians. Thus, for very small angles $\theta$, we have, to a very good approximation, sin $\theta = \theta$. But the distance from P to the $x$ axis (the sine of $\theta$) is always smaller than the arc length from P to the $x$ axis (the angle $\theta$) and the difference between these two lengths increases as $\theta$ increases. In other words, placing sin $\theta = \theta$ is a gross error for large angles; thus, for $\theta = \pi/2$ (90°) the sine is 1 whereas the angle $= \pi/2 = 3.1416/2 = 1.5708$. To obtain a better formula for sin $\theta$ we introduce another term and write sin $\theta = \theta - \theta^3/6$. But this expression reduces the formula too much, which gives values for the sine of $\theta$ that are too small. To improve the formula we add a positive term to obtain sin $\theta = \theta - \theta^3/6 + \theta^5/120 - \cdots$. We can go on adding as many additional terms as we please, alternating from "+" to "−" and using only odd powers of $\theta$ to ensure that the sine of a negative angle is negative. This alternating series for sin $\theta$ gives highly accurate values even if only four or five terms are used in the series because the additional terms have larger and larger denominators which make those terms smaller and smaller. Thus, the fourth term, which comes in with a minus sign, is $-\theta^7/5040$, which is very small indeed, even if $\theta = \pi/3 = 1.0472$ or 60°.

The infinite series for the cosine is also an alternating series of powers of $\theta$, but the powers are all even and the series begins with 1 because for angles close to zero the cosine is 1 since the cosine is the distance from P to the $y$ axis; this quantity is just the unit radius when P is very close to the $x$ axis ($\theta$ is very small). We therefore have $\cos \theta = 1 - \theta^2/2 + \theta^4/24 - \theta^6/720 + \cdots$.

We emphasize that everything we have presented in this chapter stems from the definition of the sine of an angle (or its cosine) so that trigonometry is nothing more than the body of the relationships we can deduce from this definition. Owing to the connection between the sine, cosine, and tangent of an angle and the geometry of the unit circle, these functions of an angle are called circular functions. In the next chapter we shall see how the great 17th century French philosopher and mathematician René Descartes developed the algebra of geometric figures of various kinds and so founded what is now called analytical geometry. This discovery led to the calculus.

## NOTES

[1] Morris Kline, *Mathematics and the Physical World.* New York: Dover, 1981, p. 11.

[2] Philip E. B. Jourdain, "The Nature of Mathematics," reprinted in James R. Newman's *The World of Mathematics.* New York: Simon & Schuster, 1956, p. 8.

[3] Hermann Weyl, "Symmetry," reprinted in James R. Newman's *The World of Mathematics.* New York: Simon & Schuster, 1956, p. 671.

[4] *Ibid.,* p. 672.

CHAPTER 7

# Analytic Geometry

*In order to seek truth, it is necessary once in the course*
*of our life to doubt as far as possible all things.*
—RENÉ DESCARTES

Until the 17th century when René Descartes combined
algebra and geometry into what we now call analytic
geometry, these two branches of mathematics had evolved
independently of each other. Geometry goes back to before
the time of the Greek mathematician Euclid and algebra
originated in ancient Egypt and Babylonia and with the
medieval Arabs who introduced the modern algebraic nomenclature. The earliest known Western treatise on algebra was
written in about 300 A.D. by the Greek Alexandrian
Diophantus, who pioneered the solutions of certain special
types of algebraic equations; this work led to the branch of
mathematics now called Diophantine analysis. This kind of
algebra deals with equations that have two unknowns and
integers as solutions.

We do not know who first proposed the idea of expressing
geometrical relationships algebraically but Descartes placed
analytical geometry on a sound foundation by introducing

the concept of the coordinate system which we used extensively in the previous chapters. The coordinate system is a kind of device that transforms an algebraic relationship into a geometrical figure or vice versa. To see how this happens we review briefly how the coordinate system forced its way into mathematics. We recall that we introduced the number system by labeling points on a line using all possible combinations of the ten basic symbols 0, 1, 2, 3, 4, 5, 6, 7, 8, 9 in groups of one, two, three, and so on, without limit. In assigning a special role to the point labeled 0 by calling it the origin (starting point) we establish the basis of a coordinate system for we then know in principle how to locate any point on the line relative to 0 if we introduce a unit of distance along the line which does not need to have anything to do with a real distance in space unless we specifically indicate that the separations between points on the line refer to real distances. To make this clear, we note that the numbers assigned to the points on the line may refer to any kind of quantity, for example, time, which we may picture as increasing in seconds from some zero (0) moment. Thus, the combination 10 of our symbols would then represent a passage of ten seconds of time. Although we may use the points on the line and the numbers we attach to them to represent any measurable entity we wish, we associate them with spatial distances for the time being because we shall be dealing with the algebra of actual geometrical figures. To do this we must go to points outside the line and, hence, sets of lines (two-dimensional coordinate systems) as we already did in the chapters on geometry and trigonometry.

We obtain a two-dimensional coordinate system if we pass another straight line through the point 0 that we chose as the origin on our first straight line. These two straight lines determine one and only one plane in space. We shall limit ourselves here to geometric figures that are formed by lines or curves that lie on this plane. In short, our coordinate system must enable us to locate any point on the plane by giving its position with respect to the lines that define this coordinate system. We do this by assigning two numbers to every point; these numbers are called the coordinates of the point. Although we define our coordinate system here in terms of two sets of straight lines (the lines in each set must be parallel to each other but the two sets must cut each other at some angle), we can also use two different sets of curves or a set of curves and a set of straight lines. Since two sets of straight lines give the simplest kind of coordinate system, we first develop the basic principles of analytic geometry in terms of such a coordinate system (called the Cartesian coordinate system) and later indicate how other coordinate systems can be used. In setting up our Cartesian coordinate system we have a choice as to the angle the two sets of parallel lines that define our coordinate system are to make with each other. Here again, we are guided by simplicity which dictates that the two sets of parallel lines cut each other at an angle of 90° (the two sets are perpendicular or normal to each other). Since the two intersecting sets of parallel lines in such a coordinate system cover the plane with rectangles (a rectangular mesh), this type of coordinate system is called a rectangular Cartesian coordinate system.

We may rotate this mesh any way we want in the plane so that, for example, one set of parallel lines runs east–west and the other set runs north–south like the lines of latitude and longitude on a map. We now choose any point where an east–west and north–south line cross and call it the origin 0 of our coordinate system. We call the east–west line through this point the $x$ axis of our coordinate system and the north–south line through it the $y$ axis. The coordinates $x, y$ of any point in the plane are the point's distances from the $y$ axis and $x$ axis, respectively. Having chosen the origin 0 and the $x$ (abscissa) and $y$ (ordinate) lines, we eliminate all other lines in our original two sets of lines and study the various algebraic representations of the $x$-$y$ coordinates of sets of points on the $x$-$y$ plane that lie on unbroken curves or lines. The essence of analytic geometry is to describe curves on the $x$-$y$ plane by giving the algebraic relationships (equations) that connect the $x$-$y$ coordinates of the points on these curves, or, conversely, to find the curve (called the locus of points by the mathematician) on the $x$-$y$ plane that pictorializes (plots or graphs) any given algebraic relationship between a set of quantities $y$ and a set of quantities $x$.

We limit ourselves here to the description of analytic geometry on a plane, which means that the figures that can be drawn and which we shall deal with obey Euclidean (flat) geometry; this is not true for the figures on a curved surface such as a sphere. Since the difference between the analytic geometry of a flat surface (a Euclidean two-dimensional manifold) and that of a curved surface can be determined simply by analyzing the algebraic formula for the distance between two neighboring points on the surface, we begin by expressing this distance formula for two such points labeled 1 and 2 on our $x$-$y$ plane in terms of the coordinates of these points. The values $x_1$, $y_1$ are the coordinates of point number 1 and the values $x_2$, $y_2$ are the coordinates of point number 2 (the subscripts 1 and 2 identify the coordinates of the two points). To make things specific we choose point 1 to be to the left of and below point 2, both of which are located in the first quadrant of the coordinate system. This means that $x_1$ is smaller than $x_2$ (written as $x_1 < x_2$) and $y_1$ is less than $y_2$ (written as $y_1 < y_2$).

We now consider the right triangle formed by the line $s$ connecting points 1 and 2, a line from point 1 drawn to the right parallel to the $x$ axis, and a third line drawn downward from point 2, parallel to the $y$ axis, until it meets the line from point 1 at the point we label P. The vertices of this right triangle are the points 1, 2, and P; the lengths of its sides are $s$ (its hypotenuse, the distance between the points 1 and 2), the distance $(x_2 - x_1)$ of P from point 1, and the distance $(y_2 - y_1)$ of P from point 2. From the theorem of Pythagoras we know that $s^2 = (x_2 - x_1)^2 + (y_2 - y_1)^2$ and the distance between the two points is the square root of the quantity $s = \sqrt{(x_2 - x_1)^2 + (y_2 - y_1)^2}$. Thus, the distance between any two points on a plane, no matter how far apart they might be spaced, is given entirely in terms of the four coordinates $x_1$, $y_1$, $x_2$, $y_2$ of the two points. This is true for two points on a sphere only if the two points are very close to each other; the larger the sphere, the greater is the allowed separation between the points for which the formula is applicable.

The simplest curve in analytic geometry is the straight line, which we discussed and described in terms of a coordinate system in our chapter on algebra. We return to it here to study more thoroughly its relationship to the coordinate system we set up than we did previously. Since all points on a plane are identical, the lines on a plane have no intrinsic distinguishing features which enable us to identify a line by its position or direction on the plane. The introduction of a coordinate system permits us to do this operation and the algebraic equation which describes any line identifies it completely and uniquely with respect to the coordinate system we have introduced. Each line is described by one and only one algebraic equation which relates the $y$ coordinate of any point on the plane to its $x$ coordinate in a unique way. Note that an infinite number of straight lines can pass through a single point on the plane but only one line can pass through two different points. Moreover, a given straight line cuts the $x$ axis just once and forms a single angle with the $x$ axis.

Thus, a straight line is completely determined if we specify any two points through which it passes or if we specify one point on the line and also the angle it makes with the $x$ axis. This information must be contained in the algebraic equation of the straight line in our coordinate system.

Consider first all the lines that pass through the origin 0; only one of these lines also passes through the point whose coordinates are $x$, $y$. The algebraic equation of the line must show this because we know that if $\theta$ is the angle the line makes with the $x$ axis, then the tangent of $\theta$ is $y/x$. Hence, we have the equation $y/x = \tan \theta$ or, on multiplying both sides by $x$, $y = x \tan \theta$. If we use the letter $a$ to represent $\tan \theta$, which is called the slope of the line, the equation of the line is $y = ax$. Since $y = 0$ for $x = 0$, the line passes through the origin 0, and if we move along the line, we rise $a$ units of length above the $x$ axis for each unit of length we advance along the $x$ axis so that $y = a$ for $x = 1$, $y = 2a$ for $x = 2$, and so on. The equation $y = 0$ describes the $x$ axis itself and $x = 0$ describes the $y$ axis.

If the line does not pass through the origin 0, it must cut the $x$ axis and $y$ axis in two points that are different from 0. The point where it cuts the $x$ axis is called the $x$ intercept of the line and the point where it cuts the $y$ axis is called the $y$ intercept of the line. We obtain the correct algebraic equation of a line that does not pass through the origin 0 by adding a constant (a quantity that does not depend on $x$) to the right-hand side of the equation $y = ax$. If we call this constant $b$, we obtain the general equation $y = ax + b$ for a straight line in our coordinate system. Note that $b$ does not change the slope of the line; it simply shifts the entire line upward or downward (if $b$ is positive, every point on the line $y = ax$ is shifted upward along the $y$ axis by $b$ units of length; if $b$ is negative, every point on the line $y = ax$ is shifted downward by $b$ units of length). Since this line intersects the $y$ axis at a point $b$ units above the origin 0 (if $b$ is positive), $b$ is called the $y$ intercept of the line. The equation $y = ax + b$ of the line tells us that this is true because if we place $x = 0$ (which means a point on the $y$ axis) in the equation, we obtain $y = b$. We obtain the $x$ intercept by placing $y = 0$ in the equation so that we have $0 = ax + b$ or $ax = -b$ or $x = -b/a$, which is the $x$ intercept.

The equation of a straight line tells us that a line is completely determined by the two constants $a$ and $b$, its slope and its $y$ intercept. No two lines can have the same slope and the same $y$ intercept. Two or more lines that have the same $a$ (slope) are parallel and any number of lines that have the same $b$ pass through the same point on the $y$ axis. That a line is completely specified by two constants is equivalent to the statement that two different points on the plane determine a straight line uniquely. Having the general equation $y = ax + b$ for any line, we can immediately find how the equation looks if it passes through two given points; the slope $a$ of such a line and its $y$ intercept $b$ must be determined entirely by the four coordinates of the two points. To see how all this works out, we derive the equation of the line that passes through the two points labeled 1 and 2 which have the coordinates $x_1, y_1$ (point 1) and $x_2, y_2$ (point 2). Since point 1 lies on the line, we have $y_1 = ax_1 + b$; for point 2 on the same line (same $a$ and $b$ values), we have $y_2 = ax_2 + b$. We call two such equations simultaneous equations for the two unknowns $a$ and $b$; finding $a$ and $b$ in terms of $x_1$, $y_1, x_2, y_2$ is called solving these equations.

We recall from Chapter 2 that we may do anything we please to an equation or to sets of equations as long as we do not destroy any of the equalities expressed by the equations. Thus, we may add equations together, subtract one from the other, multiply one by the other, or divide one by the other without violating any algebraic rules. With this reassurance we may subtract the second equation for point 2 from the first equation for point 1 to obtain $y_1 - y_2$ $= ax_1 - ax_2 + b - b = ax_1 - ax_2 = a(x_1 - x_2)$. We now divide both sides of this equation (the term on the extreme left and that on the extreme right) by $(x_1 - x_2)$ so that we have $(y_1 - y_2)/(x_1 - x_2) = a$. Thus, the slope $a$ of a line passing through the two points is $(y_1 - y_2)/(x_1 - x_2)$. The equation of the line must be $y = [(y_1 - y_2)/(x_1 - x_2)]x + b$ where we have replaced $a$ by the slope expressed in terms of the four coordinates of our two points.

To find the value for $b$ we note that this equation must hold for every point on the line and, therefore, for the point 1, which means it must hold for $y = y_1$ and $x = x_1$. Hence, we have $y_1 = [(y_1 - y_2)/(x_1 - x_2)](x_1) + b$ so that, after subtracting $[(y_1 - y_2)/(x_1 - x_2)](x_1)$ from each side,

$$y_1 - [(y_1 - y_2)/(x_1 - x_2)](x_1) = b$$

If we multiply both sides of this equation by $(x_1 - x_2)$ we obtain

$$(x_1 - x_2)y_1 - [(y_1 - y_2)/(x_1 - x_2)](x_1)(x_1 - x_2) = b(x_1 - x_2)$$

or

$$(x_1 - x_2)(y_1) - (y_1 - y_2)(x_1) = b(x_1 - x_2)$$

After multiplying out the left-hand side we obtain

$$x_1 y_1 - x_2 y_1 - x_1 y_1 + y_2 x_1 = b(x_1 - x_2)$$

or

$$y_2 x_1 - x_2 y_1 = b(x_1 - x_2)$$

After dividing both sides by $(x_1 - x_2)$ we finally obtain $(y_2 x_1 - x_2 y_1)/(x_1 - x_2) = b[(x_1 - x_2)/(x_1 - x_2)]$ so that $b = (y_2 x_1 - x_2 y_1)/(x_1 - x_2)$. The equation for the line that passes through the two points 1 and 2 is thus

$$y = [(y_1 - y_2)/(x_1 - x_2)](x) + (y_2 x_1 - x_2 y_1)/(x_1 - x_2)$$

If point 1 has the coordinates $x_1 = 3$, $y_1 = 5$ and point 2 has the coordinates $x_2 = 4$, $y_2 = 7$, the equation of the line is

$$y = \frac{5-7}{3-4}x + \frac{(7)(3) - (4)(5)}{3-4}$$

$$= \frac{-2}{-1}x + \frac{21-20}{-1}$$

$$= 2x + \frac{1}{-1} = 2x - 1$$

The slope of this line is 2 and its $y$ intercept is $-1$ which means that it cuts the $y$ axis one unit of distance below (owing to the minus sign) the origin 0. Its $x$ intercept is obtained by placing $y = 0$ so that $2x_{\text{intercept}} - 1 = 0$, or $2x_{\text{intercept}} = 1$; hence, $x_{\text{intercept}} = 1/2$.

We have gone through this process in this very long-winded way (what mathematicians call a hammer-and-tongs process) to illustrate in detail how one uses standard algebraic (actually arithmetic) rules to solve a problem which is essentially geometric, but is expressed algebraically. But we can obtain the same result without arithmetic or algebra; instead we use the symmetry that the equation of the line must have with respect to the two points through which it passes. Here we proceed the way a detective does when he uses all the available clues to solve a crime. The equation of a line $y = ax + b$ gives us the two clues we need to find $a$ and $b$ if the line passes through the two points $x_1$, $y_1$ and $x_2$, $y_2$. Since $y$ is a distance, each term on the right-hand side is also a distance. But this means that $a$ cannot be a distance since $x$ is already a distance; $a$ must therefore be a pure (dimensionless) number. However, $b$ must be a distance. Thus, we seek two quantities, one a pure number—the slope of the line—and the other a distance—the $y$ intercept. For our second clue we turn to symmetry. The equation of the line must contain the coordinates of the two points in exactly the same way because it must be symmetrical with respect to the two points; if we interchange the two points, the line is still the same. With the aid of these two clues we can now write down the equation at once.

First, we deduce the expression for the slope $a$ of the line. If the line were exactly horizontal (parallel to the $x$ axis), its slope would be zero. Therefore, the slope $a$ must contain a term which equals 0 when the line is parallel to the $x$ axis. The two points 1 and 2 through which the line would then pass would be equally high above the $x$ axis which means that $y_1$ and $y_2$ (their ordinates) would be equal and $y_2 - y_1$ would equal 0. We therefore introduce $(y_2 - y_1)$ as part of the expression for $a$, but this equation cannot be the entire expression because $(y_2 - y_1)$ is a distance whereas $a$ is a dimensionless quantity, which we obtain by dividing $(y_2 - y_1)$ by another distance which must not contain $y_2$ or $y_1$. The only other quantities we have with which to construct a distance are $x_1$ and $x_2$. We can obtain such a distance by combining these two distances as a sum or a difference. We are immediately guided to the difference as the correct combination if we note that if the two points were equally far from the $y$ axis so that $x_1$ were equal to $x_2$, the line passing through them would be exactly vertical (parallel to the $y$ axis) and would therefore be infinitely steep because its slope would be infinite. This means that $a$ must depend on $x_1$ and $x_2$ in such a way that it becomes infinite when $x_1 = x_2$ or when $x_2 - x_1 = 0$. The two conditions we have deduced for $a$ and the requirement that it be dimensionless are fulfilled for $a = (y_2 - y_1)/(x_2 - x_1)$, for $a = 0$ for $y_2 = y_1$, and infinite for $x_2 = x_1$ (an algebraic expression becomes infinite if its denominator becomes zero). Note that the expression for $a$ remains the same if we interchange $y_2$ and $y_1$ and, at the same time, interchange $x_2$ and $x_1$, which is equivalent to interchanging the two points.

Finding the value for $b$ (a distance) is simpler than finding the value for $a$. We note that if the two points lie on the same vertical line, the $y$ intercept is infinitely far away because the line is parallel to the $y$ axis (parallel lines intersect at infinity). Since $x_1 = x_2$ or $x_1 - x_2 = 0$ for two points on a vertical line, the difference $(x_1 - x_2)$ must be in the denominator of the algebraic expression for $b$ to make $b$ infinite when the line is parallel to the $y$ axis. Since $b$ is a distance (a length), the numerator of the expression for $b$ must be a length squared (a product of two lengths) if the entire expression is to be a length as required by the equation of the line. Since we have to use $(x_1 - x_2)$, a length, as the denominator of the expression for $b$, we have to use a length squared as the numerator to obtain (length)$^2$/length = length. The only symmetric combination of our coordinates that has the spatial dimensions of (length)$^2$ and, when combined with the denominator $(x_1 - x_2)$, gives the correct $y$ intercept, namely $y_1$ or $y_2$, when the line is parallel to the $x$ axis (horizontal), is $y_2x_1 - x_2y_1$. Hence, the correct expression for $b$ is $(y_2x_1 - x_2y_1)/(x_1 - x_2)$, as derived previously using algebra.

We have spent considerable time explaining the use of symmetry and dimensional concepts to solve problems because, when properly applied, these concepts give a deep insight into problems and indicate the correct paths to their solutions. They have been used by the greatest mathematicians and physicists to unravel mysteries that appeared impenetrable to others.

An interesting and important application of the equation of a straight line is to the solution of simultaneous equations containing two unknowns. The equations we discussed in our chapter on algebra are equations with one unknown $x$. If the problem we have to solve contains two unknowns, which we call $x$ and $y$, we must have two equations involving the two unknowns before we can find the two unknowns, which means we must solve the two equations together. The numbers $x$ and $y$ which we obtain when we solve the two equations may refer to any kinds of quantities which need not be related to distances in any way, but we can still use the equations of straight lines and a coordinate system to solve such equations. As an example, we suppose that, having been told that Mary and her brother John together have a certain amount of money $P$, at some moment in time, and that Mary's money plus twice John's money at that time equals $3P$, we are then asked to determine the amount of money held by each sibling at that moment. If $M$ and $J$ stand for the amount of money they have separately and $P$ is given, we can set up two equations for the two unknowns $M$ and $J$. These equations are $M + J = P$ and $M + 2J = 3P$. If we set up a certain coordinate system in which $M$ is plotted along the ordinate axis (the $y$ axis) and $J$ is plotted along the abscissa (the $x$ axis), each equation represents a straight line in this coordinate system, which is obvious, based on our previous discussions if we write the two equations as $M = -J + P$ and $M = -2J + 3P$. The plane defined by the $M$, $J$ coordinate system is an imaginary plane, each point of which may represent the amount of money, $M$, held by Mary and the amount of money, $J$, possessed by John. If these amounts vary from day to day or week to week, the point which tells us how much money each sibling possesses changes accordingly ($M$ is the ordinate of the point and $J$ is its abscissa; note that the length of the line from the origin 0 to the point has no meaning in this case). Only one point among all the points on this imaginary plane has $M$ and $J$ values that solve the problem; this point must therefore lie on both lines given by the two equations for $M$ described above. The only point that lies on two different straight lines is the point of intersection of the lines; since each line given above has a different slope ($-1$ and $-2$, the coefficients of $J$), they are not parallel and therefore must intersect in a point. The $M$ and $J$ coordinates of that point solve the problem. This means that we can solve the given simultaneous equations graphically by drawing the two straight lines which are represented by the equations and finding their common point of intersection.

We return now to the geometrical plane and introduce a concept concerning coordinate systems which is of great importance in mathematics and physics. To arrive at this concept we picture two different mathematicians contemplating the same plane, each with his own rectangular Cartesian coordinate system. How will their descriptions of points, lines, etc., on this plane differ? If one of the mathematicians (whom we shall call No. 1) picks some point 0 as his origin and chooses his $x$ and $y$ axes (at right angles to each other) to be oriented in some direction of his choice, the second mathematician (whom we shall call No. 2) may choose a different origin which we may describe as 0', a different orientation for his axes, or both, for his coordinate system. Owing to these differences, the coordinates assigned to a point on the plane by mathematician 1 differ from those assigned to the same point by mathematician 2. These differences are easy to describe algebraically if we treat them separately. We consider the shift in the origin from 0 to 0' first. If P is a point on the plane which, in the coordinate system of mathematician 1, has the coordinates $x$ and $y$ (it is $x$ units of distance from the $y$ axis and $y$ units of distance from the $x$ axis), the coordinates $x', y'$ in the coordinate system of mathematician 2 are different from $x$ and $y$ because they are distances measured from axes that are shifted from the axes of the first coordinate system. If the origin 0' is $k$ units of distance to the right of 0, then the $x'$ coordinate of a point is smaller by exactly $k$ units of length than the $x$ value of that point. Thus, $x' = x - k$ for every point. If 0' is $h$ units of length above 0, then the $y'$ value of each point is smaller than its $y$ value by the amount $h$ and we have $y' = y - h$. These two equations for $x'$ and $y'$ define what mathematicians call a transformation of coordinates (which is, in this case, a translation of the origin from 0 to 0'). This transformation does not alter any intrinsic features of the geometry of the plane. Thus, it leaves distances between points and angles between lines unchanged.

We consider now the effect of a difference between the orientation of the coordinate axes of 0 and that of the axes of 0' only (no translation) on the coordinates of a point. In this situation the origin 0 and 0' coincide but the direction of the $x'$ axis differs from that of the $x$ axis and the direction of the $y'$ axis differs from that of the $y$ axis by the same amount. As a specific example, we suppose that mathematician 2 chooses an $x$ axis that is tilted (rotated) counterclockwise by 30° away from the $x$ axis of mathematician 1. This rotation does not alter the distance of any point from the common origin 0 and 0' of the coordinate system but the $x', y'$ coordinates of the point (the distances of the point from the $x'$ and $y'$ axes) are different from $x, y$ because these distances depend on the angle that the line from the origin makes with the $x'$ axis. Hence, the coordinates $x', y'$ are different from the coordinates $x, y$. The rotation of the coordinate system changes the coordinates of the points on the plane but it does not change the distances between points or the angles between lines. If certain mathematical or physical entities do not change when we alter our coordinate system in any way, we say that such entities are invariant to a transformation of coordinates. This general concept, called the principle of invariance, plays a very important role in mathematics and physics. In the simple example of analytic geometry on a plane, the distances between points and the angles between lines are invariant.

In addition to using straight lines as the basis for our coordinate system, we can use sets of curves or curves and straight lines. The simplest and most useful coordinate system of this type is the polar coordinate system (discussed in an earlier chapter) which consists of a system of concentric circles with the origin as the center of these circles and a system of straight lines radiating out from the center (origin). The position of each point P on the plane is completely specified by the radius $r$ of the circle that passes through it and the angle $\theta$ that the radius $r$ from the origin 0 to the point P makes with the $x$ axis. Thus, the two coordinates of the point P in a polar coordinate system are its distance $r$ from the origin 0 and the angle $\theta$. Note that only one of these coordinates is a distance; the other coordinate is an angle. The longitudes and latitudes of points on the surface of the Earth are essentially polar coordinates with the origin of this polar coordinate system at the North Pole. One cannot draw straight lines on the Earth's surface but the great circles of longitude on the Earth that all radiate from the North Pole are equivalent to straight lines since they are the shortest distances between points lying on them; the circles of latitude are equivalent to circles on a plane. If the Earth were extremely large, longitude and latitude would correspond almost exactly to the $\theta, r$ polar coordinates on the plane.

We devote the rest of our discussion of analytic geometry to conic sections, which play a remarkable role in the theory of the orbits of bodies pulling on each other gravitationally or electrostatically. If we spin two intersecting lines on a plane about a line (in the plane) that bisects their angle of intersection, we obtain two right circular cones with the point of intersection as their common vertex; the two cones stand head on head (vertex to vertex) and the angle bisector is their altitude. These cones are three-dimensional figures so that any plane passing through them in any direction cuts them in a curve or curves of some sort (called conic sections) that depend on the orientation of the cutting plane with respect to the altitude of the cones. If the cutting plane is exactly perpendicular to the altitude, the curve of intersection of the plane and the cone (the conic section) is a circle. However, if the cutting plane is not perpendicular to the altitude, and not parallel to either of the intersecting lines (which are called generators) that were spun to generate the cone, the curve of intersection is an ellipse. If the cutting plane is parallel to either generator, it cuts the cone in a parabola. Finally, if the cutting plane is tilted so that it cuts both cones, two identical curves of intersection called the branches of a hyperbola are produced.

Although we have introduced the circle, ellipse, parabola, and hyperbola as the intersections of planes and a three-dimensional figure (cone), we shall not refer to the cone any more and treat these conic sections as figures (curves) on a plane in their own right and see how we are to describe them

algebraically. Following the dictates of simplicity, we introduce a rectangular Cartesian coordinate system. As we shall
see, the various conic sections can be described algebraically
by quadratic algebraic equations (second-degree equations)
in two unknowns, which we may take as the $x$ and $y$ coordinates of points on the plane of our $x$–$y$ coordinate system.
The general equation of this sort is $ay^2 + bxy + cx^2 + dy
+ ex + f = 0$, where $a, b, c, d, e, f$ are numerical coefficients
of any kind except that we exclude complex numbers. Since
we may divide every term in this equation by $a$ without
altering it since 0 divided by $a$ is 0, the general second-degree
equation of conic sections can be written as $y^2 + bxy + cx^2
+ dy + ex + f = 0$. By choosing the coefficients $b, c, d, e, f$
appropriately, we obtain the equation of any conic section
we please. If we place $b = d = e = 0$ and $c = 1$ we obtain
the equation $y^2 + x^2 + f = 0$, which is the equation of a circle
whose radius squared is $-f$ ($f$ must be taken negative to
obtain a circle), and whose center is at the origin 0. By making
appropriate choices of the coefficients $d$ and $e$ (different from
0) and keeping $c = 1$ and $b = 0$, we can get the equation of
a circle whose center is any point in the plane.

   Since the ellipse is the most important of all the conic
sections in the study of the orbits of planets, we discuss these
curves in more detail, noting that the circle is a special case
of an ellipse or, put differently, the ellipse is a symmetrical
distortion of a circle. To see how we must change a circle to
obtain an ellipse, we start with a circle of radius $a$ whose
center is at the origin 0. We now construct an ellipse which
lies within the circle and touches the circle on the $x$ axis $a$
units of length to the right of 0 and also $a$ units of length on
the $x$ axis to the left of 0. In other words, both the circle and
the ellipse cut the $x$ axis at the point $x = a$, $x = -a$ ($y$ is 0
at both of these points for the circle and the ellipse). Now

the arc of the ellipse above the $x$ axis (for positive $y$ values) falls below that of the circle, departing from the circle uniformly, until the arc of the ellipse is at a maximum distance below that of the circle for $x = 0$. In other words, the separation between the two arcs is greatest for their $y$ intercepts. If we call the $y$ intercept of the ellipse $b$, then $a - b$ is the largest separation between any two points, one on the circle and one on the ellipse, with the same $x$ value. We summarize this idea as follows: The $y$ values (ordinates) of points on the ellipse increase uniformly from 0 for $x = a$ to $b$ for $x = 0$ with $b$ less than $a$ ($b < a$); the $y$ values of the points on the ellipse then decrease uniformly from $b$ to 0 as the $x$ values of the points decrease (become more negative) from 0 to $-a$. The length $a$ (the largest distance of a point on the ellipse from the origin 0) is called the semimajor axis of the ellipse. The length $b$ (the smallest distance of a point on the ellipse from the origin 0) is called the semiminor axis of the ellipse; $2a$ is the major axis and $2b$ is the minor axis of the ellipse.

Just as all the geometrical properties of a circle are determined by its radius (a single distance), so are all the geometrical properties of an ellipse determined by two distances: its semimajor axis $a$ and its semiminor axis $b$. We say that the circle has a single geometrical degree of freedom, whereas the ellipse has two such degrees of freedom. The semimajor axis $a$ determines the size of the ellipse, whereas $b$ determines its shape. The larger $b$ is for a given value of $a$, the rounder the ellipse, and the smaller $b$ is for a given value of $a$, the flatter the ellipse. If $b = a$, the ellipse is a circle and if $b = 0$, the ellipse is a straight line. Another interesting geometrical property of an ellipse, whose semimajor and semiminor axes are $a$ and $b$ respectively, is that its geometrical properties are a kind of average (a geometrical average) of the geometrical properties of a circle with radius $a$ and a circle with radius $b$. We began our discussion of the ellipse by considering a circle of radius $a$ inside which an ellipse is inscribed which just touches the circle at the two points ($x = a, y = 0$) and ($x = -a, y = 0$). But we could just as well have started by considering a circle of radius $b$ which lies inside our ellipse (the inscribed circle) which just touches our ellipse at the two points ($x = 0, y = b$) ($x = 0, y = -b$). The arc of our ellipse lies between the circumferences of these two circles; it is inscribed in the larger one and circumscribes the smaller one. From these considerations we see that the ellipse is symmetrical with respect to the $y$ and $x$ axes. The shape of the ellipse to the right of the $y$ axis (for positive $x$ values) is the mirror image of its shape to the left of the $y$ axis (for negative $x$ values); its shape above the $x$ axis (for positive $y$ values) is the mirror image of its shape below the $x$-axis (for negative $y$ values).

We now introduce another way of describing the shape of an ellipse which is based on a quantity called the eccentricity of the ellipse: a pure number (no spatial dimensions) which is a measure of the amount by which the ellipse departs or differs from a circle. The word "eccentricity" is appropriate for this measure since it is derived from "eccentric" which means off-center, and the ellipse, in a sense, is not a centered curve like a circle. The "eccentricity" of an ellipse can be defined in various ways, but, however we define it, it must vary from 0 to 1 as we go from a circle (eccentricity = 0) to a straight line (eccentricity = 1). A quantity which varies just this way is $(1 - b/a)$ where $b$ is the semiminor axis and $a$ is the semimajor axis of the ellipse we are discussing. By calling this quantity $e$ (for eccentricity), we see that if $b = a$ (a circle,

since the semiminor and semimajor axes are equal), then
$e = 1 - a/a = 1 - 1 = 0$. If $b = 0$, the curve is a straight line
and $e = 1 - b/a = 1 - 0 = 1$.

Having discussed the geometric properties of the ellipse
from the point of view of general symmetry principles, we
now show that it is described by a general algebraic equation
which relates the $x$, $y$ coordinates of any point on the ellipse
to the basic lengths $a$ and $b$ (its semimajor and semiminor
axes respectively). We shall be guided by the equation of the
circle. If $x$, $y$ are the coordinates of a point on the circumference of a circle of radius $a$, the equation of the circle is
$x^2 + y^2 = a^2$ (the theorem of Pythagoras). If we divide both
sides of this equation by $a^2$, we have for the same circle
$(x/a)^2 + (y/a)^2 = 1$. Can we describe an ellipse by a similar
equation? Yes, if we note that while the equation of the circle
deals with only the three lengths $x$, $y$, and $a$, the equation of
the ellipse must involve the four lengths $x$, $y$, $a$, and $b$.
Returning to the equation of the circle $(x/a)^2 + (y/a)^2 = 1$
in which our ellipse is inscribed, we see that this equation
gives the correct coordinates for two points on the ellipse—the
two points where the circle and our inscribed ellipse touch
on the $x$ axis. Since $y = 0$ for these two points on the $x$ axis,
their $x$ coordinates are obtained from the equation of the
circle by placing $y = 0$ in that equation. We then have
$(x/a)^2 + 0 = 1$ or $x/a = \pm\sqrt{1} = \pm1$ so that $x = \pm a$. Since
these are two points on the ellipse, the first term $(x/a)^2$ in
the equation for the circle must also be present in the equation
of the ellipse. The equation of the ellipse is therefore $(x/a)^2$
$+? = 1$, where the question mark indicates that we still have
to determine the second term in the ellipse's equation. We
apply symmetry principles to deduce this term. The symmetry
of the ellipse with respect to the $x$ and $y$ axes tells us that $y$
must appear in the equation in exactly the same way as $x$

does so that the second term must be of the form $(y/?)^2$ where the question mark stands for a distance; the only distance (length) other than $a$ that characterizes the ellipse is its semiminor axis $b$. The second term is therefore $(y/b)^2$ so that the equation of the ellipse is $(x/a)^2 + (y/b)^2 = 1$ which gives us the relationship of the $y$ coordinate of every point on the ellipse to its $x$ coordinate. If we multiply this equation by $b^2$, we obtain $(b/a)^2 x^2 + y^2 = b^2$ or $y^2 + (b/a)^2 x^2 - b^2 = 0$, which is a special case of the general quadratic binomial.

We note a few more interesting features of the ellipse which are useful in the study of the orbits of planets and satellites. The area of an ellipse can be deduced from general dimensional and symmetry considerations. Since area is the square of a length (dimensionally), the formula of the area of an ellipse must be some pure number multiplied by the square of some characteristic length or by the product of two such lengths. A circle has only a single characteristic length—its radius $r$—so that its area is $\pi r^2$; but an ellipse has two characteristic lengths—its semimajor axis $a$ and its semiminor axis $b$ and these two lengths must appear in the formula for the area in exactly the same way. The only possibility for the area formula for an ellipse is therefore $\pi ab$; it has the right symmetry characteristics with respect to $a$ and $b$ since the formula remains the same if we interchange $a$ and $b$. Moreover, if $a = b$, we have a circle of radius $a$ and the area formula becomes $\pi a^2$, as it should for a circle. Thus, the formula $\pi ab$ is exactly what we want for the area of an ellipse with semimajor and semiminor axes $a$ and $b$, respectively. Note that if $b = 0$, the area is zero, as it should be for a straight line; the area of the ellipse thus goes from 0 to the area of a circle as $b$ goes from 0 to $a$. Since $b = a\sqrt{1 - e^2}$, the area of the ellipse can also be written as $\pi a^2\sqrt{1 - e^2}$ which shows how the area depends on the eccentricity $e$ of the ellipse. The area of the ellipse along which a planet moves around the sun plays an important role in the dynamics of the planet's motion, as was first discovered in the early 17th century by Johannes Kepler.

We now introduce a different way of expressing the equation of an ellipse, which is used extensively in planetary orbit theory. To this end we study the change in the distance of a point on the ellipse from the origin 0 if 0 is shifted to the left on the $x$ axis by an amount $d$ to a new position which we call 0'. If $d$ is zero (not shifted), the distance from 0 of a point on the ellipse decreases from $a$ (the semimajor axis) for the point of the ellipse on the $x$ axis to the distance $b$ (the semiminor axis) for the point of the ellipse on the $y$ axis; it then increases to $a$ again for the left point of the ellipse on the $x$ axis. In other words, the distances from 0 of two left and right symmetrically situated points vary from $a$ to $b$. If 0 is now shifted to the left of its original position by the amount $d$, on the $x$ axis, its distance from the point to the right where the ellipse cuts the $x$ axis (the right tip of the major axis) is $(a + d)$ and therefore larger than $a$ and its distance from the left tip of the major axis along the $x$ axis is $(a - d)$ and therefore smaller than $a$. Hence, the distance of 0' (0 in its shifted position) from points on the ellipse decreases from $(a + d)$ for the point on the right tip of the major axis to $(a - d)$ for the point on the left tip of the major axis. Since this distance decreases from a value larger than $a$ to one smaller than $a$ for points on the ellipse going from the extreme right to the extreme left, the distance of just one of these points from 0' exactly equals $a$. Just where this point is positioned on the upper arc of the ellipse depends on the distance $d$; the bigger the value for $d$ (the further we shift 0 to the left), the closer this point is to the upper tip of the semiminor axis. It coincides with this tip for a definite displacement $d$ of the origin 0 to the left (for a definite position $0'$). The same is true if we shift the origin 0 to the right by the proper distance $d$. We thus discover two points F and F′ on the major axis of the ellipse (equidistant from the center 0) which are at exactly the distances $a$ from the tip of the minor axis. These points, F and F′, are called the foci of the ellipse; they are equidistant $(a)$ from the tip of the minor axis, equidistant $(d)$ from the center 0 of the ellipse, and equidistant $(a - d)$ from the two tips of the major axis.

All that is left to be done in this analysis is to find the distance $d$ of each focus from the center 0 of the ellipse; we do this using the same kind of general analysis as we have employed previously in this chapter. As the ellipse gets flatter, the semiminor axis $b$ gets smaller and the two foci F and F′ have to be shifted a greater distance $d$ from 0 if each of their distances from the tip of the minor axis $b$ is to equal $a$. In other words, the flatter the ellipse, or, put differently, the larger the eccentricity, the greater is the value for $d$. If the ellipse were flat (a straight line), the eccentricity $e$ would be 1, and F and F′ would be at the left- and right-hand tips of the major axis, and $d$, of course, would equal $a$. On the other hand, if $b$ equaled $a$, the ellipse would be a circle, the eccentricity $e$ would be zero, the two foci F and F′ would coincide with the center 0, and $d$ would equal zero. From this analysis we deduce that for any ellipse of eccentricity $e$ and semimajor axis $a$, the distance $d$ equals $ea$ (the product of the eccentricity $e$ and the semimajor axis $a$). This product gives us exactly the value we want for $d$: it goes from $a$ for $e = 1$ (a straight line) to 0 for $e = 0$ (a circle). Note that if $d = ea$, the distance of each focus F and F′ from the tip of the semimajor axis $a$ on which it lies is $(a - ea)$ or $a(1 - e)$. We shall illustrate these features numerically using simple arithmetic when we apply them to the Earth's orbit around the sun.

Since the two distances of F and F' from the tip of the semiminor axis of the ellipse *ea* equal *a*, the sum of these distances equals 2*a*, which is just the length of the major axis. We now consider a point P on the ellipse either slightly to the left or to the right of the point B at the tip of the semiminor axis. If P is to the right of B, its distance from F, the focus of the ellipse to the left of 0, is slightly larger than *a* and the distance of F' from P is slightly smaller than *a*. The amount by which the distance from F to P increases, however, exactly equals the amount by which the distance of F' from P decreases so that the sum of these two distances is still 2*a*. This is true for every point P on the ellipse. We may therefore describe an ellipse as a curve consisting of all points the sum of whose distances from two fixed points F and F' equals 2*a*, where 2*a* is the length of the curve's diameter that passes through F and F'. With this geometric description of the ellipse we now have a simple way of constructing an ellipse of any size and eccentricity we please: Take a string of any length *L* and pin down the two ends at two different points F and F', separated by any distance smaller than *L*, and, holding a pencil tautly against the loop of the string, trace out a curve. The curve is an ellipse. The length of the ellipse's major axis is *L* and its two foci are F and F'. The farther apart F and F' are shifted, compared to *L*, the flatter is the ellipse (the larger its eccentricity); as F and F' approach each other, the ellipse becomes more circular, becoming a circle when F and F' coincide at the center of the line.

Astronomers simplify their mathematical description of the orbit of a planet (the Earth, for example) by using a polar coordinate system whose origin is at the sun, which occupies one of the foci F or F' of the elliptical orbit. It does not matter at which focus one puts the sun (the other focus is just a point in space); the orbits of all the planets, asteroids, and comets then have this focus in common. Although the orbit of each planet lies in its own plane, and the planes are tilted with respect to each other, all these planes pass through the common focus (the sun). With the origin of our coordinate system at the sun, we do not use Cartesian coordinates $x$ and $y$ but polar coordinates $r$ and $\theta$, where $r$ (called the radius vector) is the distance of the planet (the Earth, for example) from the sun and $\theta$ is the angle $r$ makes with the major axis of the orbit at the sun (at the focus F). As the planet moves from the far end of the major axis of its orbit (its maximum distance from the sun, which is called the planet's aphelion) to the near end of the major axis (its minimum distance from the sun, which is called its perihelion), the angle $\theta$ increases from 0 to 180°, and the cosine of this angle goes from 1 to $-1$. The distance $r$ of the planet from the sun goes from its maximum value $a(1 + e)$ where $a$ is the semimajor axis of the orbit to $a(1 - e)$, its minimum value. Since the mean distance of the planet from the sun is the sum of the $r_{\text{maximum}}$ and $r_{\text{minimum}}$ divided by 2, we find

$$r_{\text{mean}} = [a(1 + e) + a(1 - e)]/2$$
$$= (a + ae + a - ae)/2 = 2a/2 = a$$

We have deduced this mean value $a$ for the perihelion and aphelion, but it applies to distances of the planet for the entire orbit. All of these features of a planet's orbit are incorporated in the following very simple polar equation of the ellipse (the equation that relates $r$ to $\theta$): $r = a(1 - e^2)/(1 - e \cos \theta)$. This is not an algebraic equation because $r$ is not equal to a polynomial involving $\theta$ but to an expression containing the cosine of $\theta$. Note that for $\theta = 0$,

$$r = a(1 - e^2)/(1 - e) = a(1 + e)$$

and for $\theta = 180°$,

$$r = a(1 - e^2)/(1 + e) = a(1 - e)$$

since $(1 - e^2) = (1 - e)(1 + e)$. By applying this equation to the earth, we find that the value for the aphelion of the earth is 96,000,000 miles $= a(1 + e)$ and the perihelion of the earth is 90,000,000 $= a(1 - e)$. Adding these two distances and dividing by 2, we obtain 93,000,000 miles for the Earth's mean distance from the sun. If we subtract the perihelion distance $a(1 - e)$ from the aphelion distance $a(1 + e)$ we obtain $[a(1 + e) - a(1 - e)] = 2ae$; if we divide this quantity by $2a$, we obtain the eccentricity $e$ of the Earth's orbit. Thus, for the Earth's orbit we have

$$e = (96,000,000 - 90,000,000)/186,000,000$$

$$= 6/186 = 1/31$$

This very small value for $e$ tells us that the Earth's orbit is very round but not quite a circle.

We now leave the ellipse and discuss, much more briefly, the other two conic sections—the parabola and the hyperbola. We recall that we obtain the parabola as a conic section by passing a plane through the cone parallel to one of the generators of the cone. This section of the cone is an open curve which does not close but which has two symmetrical branches, one on each side of the axis (which we take as the $x$ axis) that go off to infinity. We may picture a parabola as being formed from an ellipse, whose major axis lies along the $x$ axis of our Cartesian coordinate system, if we keep the focus F of the ellipse fixed and simply move the center of the ellipse off to infinity. When we do this, the ellipse becomes larger and larger and its eccentricity, which is the ratio of the distance of the center of the ellipse from the focus F to the semimajor axis for each successively larger ellipse, increases. If 0 marks the center of any of these ellipses, the eccentricity is F0/$a$, where $a$ is the semimajor axis of the ellipse. As the center 0 moves off to infinity, both F0 and $a$ become infinite so that they become equal in size. Hence, the ratio F0/$a$ (the eccentricity) approaches the value 1. We may therefore describe a parabola as a conic section whose eccentricity is 1.

We can describe the parabola best by using the polar equation $r = a(1 - e^2)/(1 - e\cos\theta)$ with the origin of the coordinate system at the focus F. However, we must replace $a(1 - e^2)$ in the numerator of this equation by the distance P above the $x$ axis, where the parabola would cut the $y$ axis if we set up a rectangular Cartesian coordinate system at F with the $y$ axis perpendicular to the axis of the parabola (the $x$ axis). We cannot keep $a(1 - e^2)$ in the numerator of the polar equation for the parabola because $a$ is infinite and $e$ is 1 for the parabola so that the numerator would become infinity times $(1 - e^2)$ = infinity times $(1 - 1)$ = infinity times 0, which has no meaning. With $a(1 - e^2)$ replaced by P and $e$ placed equal to 1 in the denominator, the polar equation of the parabola becomes $r = P/(1 - \cos\theta)$, where $r$ and $\theta$ have the meanings introduced previously ($r$ is the distance from F to any point on the parabola and $\theta$ is the angle $r$ makes at F with the $x$ axis).

The parabolic shape plays a very important role in the design of the mirror in reflecting telescopes. When light passes through a single lens, each surface of which is perfectly spherical, the rays that are brought to a focus on the other side of the lens to form an image, do not come to a perfect focus. The image is fuzzy. This fuzziness, called lens aberration, is produced by two physical properties of spherical lenses: spherical aberration and chromatic aberration. Spherical aberration is produced by a spherical lens (glass) surface because parallel rays of light passing through the center of the lens are bent (refracted) less than those passing through the lens near its edge so that these edge rays are brought to a focus closer to the back surface of the lens than the center rays; the image thus produced is fuzzy. Chromatic aberration is produced because red light travels faster in glass than does blue light so that the red rays are bent (refracted) less than the blue rays and the image produced is therefore fuzzy and is surrounded by a halo of colors. Lens designers correct these aberrations in a good camera lens, for example, by using combinations of spherical lenses, consisting of different kinds of glass with different spherical surfaces instead of a single spherical lens. These lenses are cemented together to form a single structure. Sir Isaac Newton, who was concerned with constructing a telescope that would produce sharp, color-free images of celestial bodies, saw that he could avoid such visual distortions by replacing the lens (called the objective) in the front end of the telescope by a curved mirror at its back end. He therefore introduced a spherical mirror to produce an image of a star, giving the world its first reflector-type telescope.

When rays of light, parallel to this radius of a concave spherical mirror, strike the surface of the mirror, they are reflected and brought to a focus at a point on the radius which is halfway between the surface of the mirror and the center of the sphere of which the mirror is a part. But the focus is fuzzy because the rays reflected from the edge of the surface of the mirror are reflected at a greater angle than those reflected from the center of the surface of the mirror. This phenomenon produces spherical aberration and, hence, a fuzzy image because the rays are not all focused at the same point. But the mirror is free of chromatic aberration because all colors in a ray are reflected in exactly the same way by a mirror. Newton knew that a spherical mirror is not satisfactory for a telescope because of its spherical aberration, but he also knew that if the shape of the mirror is parabolic, all spherical aberration is eliminated. All reflectors are now constructed with parabolic mirrors so that they are free of spherical and chromatic aberrations; if rays of white light parallel to the axis of the telescope (the $x$ axis) strike the concave reflecting parabolic surface of the mirror (a paraboloid), they are all reflected to the point F (the focus) on the axis of the mirror.

We note one more geometric feature of the parabola which follows from its polar equation. If the angle $\theta$ in that equation is 90°, the cos $\theta$ term in the denominator is 0 and $r$, the distance of the focus (our origin) from the point on the parabola for $\theta = 90°$, is $P$. For $\theta = 180°$ the parabola cuts the $x$ axis. The distance $r$ of that intersection point is $P/2$ since cos 180° = −1 and the equation for $r$ becomes

$$r = P/(1 - \cos \theta) = P/[1 - (-1)] = P/2$$

We now draw a line perpendicular to the axis of the parabola (the $x$ axis) which cuts this axis at a point which is exactly $P$ units of distance to the left of the focus F. We now show that the perpendicular distance of any point on the parabola from this line exactly equals the distance $r$ of the point from the focus F. As we have seen, $r = P$ for the point on the parabola for which $\theta = 90°$ (the point directly above the focus F). But this is exactly the horizontal distance of the vertical line to the left of F and therefore its distance from the $\theta = 90°$ point. Hence, this point is $P$ units of distance from F and from the vertical line. The distance from the vertical line of any point on the parabola to the right of the point $\theta = 90°$ is $P + r \cos \theta$ where $\theta$ is the angle coordinate of the point and $r$ is its distance from F. But $r = p/(1 - \cos \theta)$ so that the expression $P + r \cos \theta$, for the distance of the point from the vertical line, is

$$p + [p/(1 - \cos \theta)] \cos \theta$$
$$= [p(1 - \cos \theta) + p \cos \theta]/(1 - \cos \theta)$$
$$= p/(1 - \cos \theta) = r$$

which proves that $r$ equals $(p + r \cos \theta)$ as we stated.

We come now to the last of the conic sections—the hyperbola. As we saw previously, this curve consists of two parts each with two open branches which we obtain if the plane cuts our two right circular cones so that two symmetric curves are produced; these curves are mirror images of each other if a mirror is placed between them. We can describe them algebraically in a rectangular Cartesian coordinate system by an equation which is similar to that of the ellipse but with a minus sign between the $x$ and $y$ terms. To this end we consider any point on the $x, y$ plane described by the equation $(x/a)^2 - (y/b)^2 = 1$ where $x$ and $y$ are the coordinates of the point in a rectangular Cartesian coordinate system and $a$ and $b$ are constants. Each such point lies on one of the two symmetrical parts which are the mirror images of each other. We may picture these parts by imagining an ellipse, with its axis along the $x$ axis of our coordinate system, sliced in half by a cut along the $y$ axis through the center 0, with each half of the bisected ellipse flipped away from the $y$ axis so that the two parts are curves that move off to infinity along the positive and negative $x$ axis. The center of these two parts, the origin of our coordinate system, is outside both parts, midway between the two points where the two parts of the hyperbola cut the $x$ axis—at the points $x = -a$, $y = 0$ and $x = a$, $y = 0$. The hyperbola also has two foci F and F', with F inside the left part at a distance $a(e - 1)$ to the left of the point $x = -a$, $y = 0$ and F' at a distance $a(e - 1)$ to the right of the point $x = a$, $y = 0$ where $e$, the eccentricity of the hyperbola, is larger than 1. We may also describe the hyperbola as consisting of all the points the differences of whose distance from the fixed points F and F′ are a constant equal to $2a$.

We obtain a special case of a hyperbola in a rectangular coordinate system if we consider all points $x$, $y$ whose coordinates obey the equation $(x)(y) = $ constant. Consider, for example, all points on the $x$, $y$ plane, the products of whose coordinates equal 6. Thus, one such point is the point having the coordinates $x = 2$, $y = 3$. This point is 2 units of distance to the right of the $y$ axis and 3 units of distance above the $x$ axis. Other such points are $(x = 3, y = 2)$, $(x = 6, y = 1)$, $(x = 1, y = 6)$, $(x = \sqrt{6}, y = \sqrt{6})$, $(x = 2, y = 3)$, and so on. If we plot all such points in the $x$, $y$ plane, we obtain a branch in the first quadrant and the mirror image of this branch in the third quadrant. If instead of 6 for the product of the coordinates, we choose another number, we obtain another parallel curve. All such curves are called equilateral hyperbolas.

# CHAPTER 8

# The Calculus and Mathematics in Science

*No nature except an extraordinary one could ever easily formulate a
theory.*

—PLATO

Of all the branches of mathematics, calculus is preeminent
in the mark of academic distinction it confers on students
who are brave enough to study it. It is required of all science
students, engineers, and architects but it remains a great
mystery to others, who either shudder at the thought of it or
view it as the most mysterious of all intellectual disciplines.
Some persons go so far as to regard calculus as a kind of
intellectual "rite of passage" or a great religious revelation.
To state that one has "taken calculus" is to place one on an
intellectual level to which few aspire, and yet, as we shall see
in this chapter, calculus is perhaps the simplest of all branches
of mathematics. Once its basic elements are understood, it
essentially explains itself. After taking calculus and discover-
ing how it simplifies intractable mathematical and physical
problems, we experience a sense of exhilaration, like first
falling in love, and we wonder at its beauty.

Despite its simplicity, the calculus is the mother of all branches of higher mathematics, of theoretical physics and chemistry, of celestial mechanics, of astrophysics, and of cosmology. Indeed, it is the most powerful and productive intellectual tool ever devised by man. Without calculus, the growth of such technologies as electricity and magnetism, radiation, electronics, atomic and nuclear physics, nuclear energy, space exploration, and most of theoretical astronomy would have been impossible. That classical mechanics developed rapidly in the 18th century from the few simple laws laid down by Isaac Newton to its present complex and beautiful mathematical edifice was no accident; it was the inevitable consequence of the discovery of the calculus in the 17th century. Like many other intellectual discoveries, which are made when their time has come and they are required for humanity's technological growth, the calculus was discovered almost simultaneously by Newton and Gottfried Wilhelm Leibniz when mathematics, physics, and technology demanded it. This discovery occurred shortly after Newton had propounded his laws of motion and the law of gravity. Newton was led to the law of gravity by his analysis of Kepler's third law of planetary motion which relates the time it takes a planet to travel around the sun (the planet's period) to the mean distance of the planet from the sun (the semimajor axis of the planet's orbit). Newton saw that he could deduce the orbit of a planet by combining his laws of motion with the law of gravity. However, he immediately ran into a difficulty while trying to accomplish this feat, because the orbit of a planet is an unchanging geometrical entity that is the same every time the planet orbits the sun. Expressed differently, the orbit is a complete, integrated entity whereas the laws of motion describe the instantaneous change of the motion of the planet in its orbit at any moment. As the planet moves around the sun, its state of motion changes continuously because its distance from the sun and its direction of motion change continuously. These changes in turn alter the force of gravity the sun exerts on the planet continuously, which in turn alters the motion of the planet again, and so on. In other words, the motion of the planet (both its direction of motion and its speed) adjusts itself continuously to meet the demands of the constantly changing force of gravity. If one knows the position of the planet relative to the sun and its motion at some initial moment, one can, from the law of gravity and the laws of motion, which give the rate of change of the planet's motion at each moment, calculate the position and motion of the planet at a very small time interval later. By piecing together all these infinitesimal changes in time over a finite interval of time, one obtains an integrated picture of the planet's orbit. Newton understood this and saw that he could deduce the orbit of a planet in this fashion if he could express the instantaneous rates of change in the position and motion of the planet at any moment in terms of the force acting on it at that moment (which also changes). He called these changes the fluxions of a quantity (in this case, the motion) and represented them by placing a dot over the changing quantity. If a planet was at some distance $r$ from the sun at some moment in time, he wrote the rate of change of $r$ as $\dot{r}$; if $v$ was the velocity of the planet at that moment, he expressed that rate of change as $\dot{v}$. Before pursuing further these ideas, which led Newton to the calculus, we consider Leibniz's great discovery which is more general than that of Newton and was directed more to abstract mathematics than to physics, as was Newton's discovery.

Leibniz was born three years after Newton and died eleven years before his British colleague. Although primarily a mathematician, Leibniz devoted much of his time to philosophy, theology, and physics. However, his approach to calculus was not from the point of view of physics but from that of the mathematical function, which we discussed briefly in our chapter on algebra. We recall that we introduced the concept of function by considering how one entity depends on another. Such dependencies are found everywhere in nature and if we do not find enough of them to suit our fancy, we can invent as many as we please. These are called functional relationships, which may be very simple or very complex. If some quantity $y$ depends on some other quantity $x$ in the sense that a given change in $x$, however infinitesimal, produces a change in $y$, we say that $y$ is a function of $x$ and write it as $y = f(x)$. This is to be understood not as a product of $f$ and $x$, which the parentheses may imply to the casual reader, but as a symbolic statement that $y$ and $x$ depend on each other. Leibniz developed the calculus from the concept of the rate of change of $y$ with respect to $x$ by considering the amount that $y$ changes if $x$ changes by an infinitesimal amount. The term "infinitesimal" is not clearly defined; Leibniz meant by infinitesimal a quantity that is smaller than any conceivable quantity and, therefore, essentially zero.

But this concept of zero change leaves the nonmathematical reader quite confused because he does not see the point of considering the change of something by a "zero" amount (the infinitesimal) because to him that means no change at all and that the whole subject is without any merit and therefore a kind of mental legerdemain. Newton and Leibniz knew, of course, that if two quantities depend on each other, then one of them changes only if the other one changes. If the change in one is zero, the change in the other is also zero. But that did not deter them from probing in depth the whole concept of infinitesimal changes of two interdependent quantities. They saw that if one goes beyond the infinitesimal changes alone and considers the ratios of those changes (the quotients), one obtains remarkable results. Newton considered such ratios as essential for his work on planetary orbits because he could deduce a planetary orbit from his laws of motion and law of gravity only if he knew the instantaneous velocity of that planet at any given moment. Knowing that the concept of instantaneous velocity is a very tricky and subtle one, he approached it very gradually. He started from the standard definition of the speed of a body as the distance a body moves in a given time divided by the time which leads to the formula $v = x/t$, where $x$ is the distance the body moves along the $x$ axis in the time $t$, starting from the origin 0. But this is certainly not the instantaneous speed if the speed is changing during the time interval $t$ which, we know, happens when we drive a car. This quotient $x/t$ gives us the average speed of the body during the time $t$ which may or may not be the instantaneous speed at any moment. To obtain the instantaneous speed at any moment, Newton saw that he had to consider the motion of the body during a very small time interval around that moment. If, at the time $t$, $\Delta t$ is a very short time interval during which the particle has moved a very small distance $\Delta x$, then the ratio $\Delta x/\Delta t$, if $\Delta t$ is short enough, is a better definition of the instantaneous speed of the particle than the quotient $x/t$. The Greek letter $\Delta$ (pronounced "delta") in front of any entity means a very small bit of that entity; under no circumstances is it to be interpreted as an algebraic coefficient multiplying the quantity following it.

By the phrase "at the time $t$" we mean following the particle during the time interval from $t$ to $t + \Delta t$ or from $t - \Delta t$ to $t$, which may be difficult in practice but is easy enough to conceive. But in contemplating the ratio $\Delta x/\Delta t$, Newton saw that, however small the quantity $\Delta t$, this ratio still gives only the average speed within a very short time interval and not the instantaneous speed. He therefore reasoned that to obtain the instantaneous speed $v$ one must allow $\Delta t$ to become zero. Mathematicians call this process passing to the limit zero and write it as

$$\lim_{\Delta t \to 0}$$

(hereafter written as $\lim_{\Delta t \to 0}$ for typesetting purposes) which means, when applied to $\Delta x/\Delta t$, the value of the quotient $\Delta x/\Delta t$ when $\Delta t$ is placed equal to zero. This procedure appears to give the meaningless, indeed nonsensical result $0/0$, which all students of elementary algebra are warned to eschew because it is clear that if $\Delta t$ goes to zero, the distance, $\Delta x$, the body moves during that time must also go to zero. But Newton was astute enough to know that although the numerator and denominator of a fraction go to zero, their ratio may have a meaningful value at the zero limit which depends on how the numerator and denominator separately go to zero. Indeed, the value of the limit ratio $\lim_{\Delta t \to 0} \Delta x/\Delta t$ depends on how $x$ depends on $t$.

That a fraction can have a perfectly good value even though its numerator and denominator both go to zero when a variable quantity in the numerator and denominator goes to some limiting value is illustrated by the fraction $(1 - x^2)/(1 - x)$. If $x$ is allowed to go to 1 in this fraction, the fraction does become the meaningless ratio $0/0$. But we avoid this embarrassment if we factor the numerator and write the fraction as $(1 - x)(1 + x)/(1 - x)$. If we now cancel the factor $(1 - x)$ in the numerator against the same factor $(1 - x)$ in the denominator, we obtain $(1 + x)$ from the fraction and its value is $(1 + 1) = 2$ when $x = 1$. This example teaches us that in allowing a variable in any algebraic expression to pass to or take on a particular value that may make the entire algebraic expression meaningless, we must first perform every possible permissible mathematical operation we can to reduce the expression to its simplest form.

Newton showed that the expression $\lim_{\Delta t \to 0} \Delta x / \Delta t$ is meaningful in general, and is, indeed, the instantaneous velocity of the body at the moment $t$. This was the beginning of what we now call the differential calculus (the calculus of the infinitesimal change of a quantity resulting from the infinitesimal changes of another quantity). Newton called the ratio $\lim_{\Delta t \to 0} \Delta x / \Delta t$ the fluxion of $x$, writing it $x$. Leibniz developed these same ideas in a somewhat different way, considering the general functional relationship between two related entities $x$ and $y$ which may or may not refer to any physical quantities. Following Leibniz we see how a mathematical entity $y$ changes when another mathematical entity $x$, on which $y$ depends, changes by an infinitesimal amount $\Delta x$. To show that $y$ depends on $x$, we express this dependence in the standard form and write $y = f(x)$ ($y$ is a function of $x$) as already stated in this chapter. We want this functional relationship to be a continuous and unbroken one over any range of $x$ we please, which may, in the extreme case, be from $x = -\infty$ to $x = \infty$. This means that for any numerical value of $x$, $y$ has a definite and unique value. For some functional relationships, the ranges of $x$ may have to be restricted because the functional relationship may be unbroken and continuous only for that restricted range of values of $x$; the functional relationship $y = 1/1 - x$ is an example of this restriction. For $x = 1$, we have $y = 1/1 - 1 = 1/0$ which is meaningless so that this function must be restricted to values of $x < 1$ (less than 1) and values of $x > 1$ (greater than 1).

Heeding these restrictions we consider the change in $y$ if $x$ changes from $x$ to $x + \Delta x$ ($\Delta x$ may be positive or negative but we always write $x + \Delta x$) so that $\Delta x$ is the infinitesimal change in $x$. Since $y = f(x)$ we write $y + \Delta y = f(x + \Delta x)$ which tells us that $\Delta y$ is the change produced in $y$ or in the function $f(x)$ when $x$ changes by the amount $\Delta x$. In other words, an infinitesimal change $\Delta x$ in $x$ results in an infinitesimal change $\Delta y$ in $y$. At this point Leibniz shifted his attention from the infinitesimal changes in $x$ and in the function $y$ of $x$ separately to the ratio of the infinitesimal change in $y$ to the infinitesimal change in $x$. In other words, he considered the fraction $\Delta y / \Delta x$ and, like Newton, asked himself how such a fraction behaves if $\Delta x$ goes to zero. We express this question symbolically as $\lim_{\Delta x \to 0} \Delta y / \Delta x = ?$ Can the question mark be replaced by a meaningful mathematical quantity? Leibniz showed that the answer is "yes" if $y$ or $f(x)$ is the right kind of function of $x$. Assuming this to be true, we introduce the nomenclature that Leibniz used and which is now employed universally by mathematicians: $\lim_{\Delta x \to 0} \Delta x / \Delta y = dy / dx$. The right-hand side $dy / dx$ is called the "derivative of the function $y$ or $f(x)$ with respect to $x$." This is also written as $df(x) / dx$ which is the instantaneous rate of change of the function $f(x)$ with respect to $x$.

How do the calculuses of Newton and Leibniz differ? They do not differ at all in principle but they do differ in emphasis. Newton was concerned with the rate of change of dynamical entities (motions of particles, etc.) with respect to time, and he placed a dot over such an entity to show the rate of change with respect to time; the dot is still used to show time rates of change. When we consider changes of various things in our own experiences or lives, we are generally concerned with changes in time, but we are concerned not with the total change of an entity over a span of time but with its rate of change, so that we deal with calculus all the time in our lives. Our patience or lack of patience is a measure of such a time rate of change.

We are interested in the time rates of change of various things that may affect our lives deeply: the rate at which our bank balance increases or decreases, the rate at which we gain or lose weight, the rate at which we age, etc. The calculus cannot be used to express these rates of change unless the time dependencies of these entities are continuous and we know them. In our daily activities, as we move from place to place, either walking or by vehicle, we are concerned with the rate of change of our distance from our destination—that is, our speed at any moment. In monitoring our motion, we are also aware that our speed is changing from moment to moment and so two different rates of change are associated with our motion—the rate of change of distance (our speed) and the rate of change of our speed, which mathematicians and physicists call acceleration.

We are concerned with many other rates of change in our daily lives which we do not monitor momentarily but become aware of over a finite time interval. As drivers, we generally know how long a tank of gasoline lasts, on the average, and therefore the time rate at which we are using our fuel. As biologists, we are interested in the rates at which various cells subdivide. As nuclear physicists, we are interested in the rates at which radioactive nuclei decay. As economists, we are interested in the rate of flow of the money supply. As employees, we are interested in the rates at which money comes to us in the form of salaries, etc. We thus see the importance of time rates of change to us, and so we must be careful to distinguish between a quantity and its rate of change. A good example of this distinction is the flow of electricity into our homes, which we start when we press a button or pull a switch; other such examples are the flow of water from a faucet and the flow of gas into our gas range when we open a valve. In these three examples we do not pay for the rates of flow of the quantities but for the total quantities themselves we receive during the time they are flowing through the wires or pipes along or through which they are conducted into our homes.

Although the three quantities cited in these examples are different, their flow rates are determined by essentially the same physical entities that are best revealed by analyzing the flow of an electric current. The current does whatever we want done (light a bulb, heat an oven, illuminate a television screen, etc.) only if it carries enough electric charge per second to the electrical device involved. The entity which we are dealing with here is the flow of electric charge; electric current is defined as the amount of electric charge that passes through any cross section of the wire in a unit time. The longer we allow the current to flow, the greater is the amount of charge brought to the device, and the larger is the quantity of light, heat, or work (depending on the nature of the device) produced by the current. We do not pay for the rate at which work is done or heat and light are produced, but for the total amount of these entities supplied to us while the current is flowing. The rate at which each device supplies what we want, which is measured by the rate it receives electric charge, is expressed in watts (a 100-watt bulb, a 10,000-watt air conditioner, etc.), which are called units of power (the rate at which the current produces energy). We pay for the total energy supplied by the current which is power multiplied by time or watts multiplied by time; our electric bills are expressed as the total number of kilowatt-hours multiplied by the price of one kilowatt-hour (a 1000-watt current flowing for one hour). This is a standard price regardless of what we want the current to do for us.

Since electric current is the flow rate of electric charges (negatively charged particles called electrons) through electric conductors (generally, metal wires) we obtain a desired current by pushing the electrons through the wire by an electric pressure called voltage; the higher the voltage for a given wire (conductor), the larger the current. Clearly, the current for a given voltage depends on the kind of wire we use (its electrical conductivity) and its dimensions; these properties of the wire determine its resistance so that the strength of a current is completely determined by the voltage across the ends of the wire (the two prongs of the electric plug) and the resistance of the wire.

This same kind of analysis applies to the flow of gas and liquids through pipes; the flow (a time rate) is defined as the amount of liquid or gas passing any cross section of the pipe per unit of time. This kind of flow or rate is of particular interest to us when it refers to the blood in our arteries or veins. Just as the flow of electric charges in wires is determined by the electric pressure on the charges and the resistance of the wires, so the blood flow is determined by the pressure (blood pressure) exerted by the heart on the blood and the resistance of veins and arteries to the flow. If this resistance increases, the blood pressure rises to keep the blood flowing at the proper rate.

We have discussed time rates of change at some length to emphasize the importance of such rates in our lives and thus to bring the calculus down to our mundane activities and strip it of its esoteric reputation. Its use and understanding are not restricted to the intellectual elite but to all who open their minds to it. The full usefulness and significance of calculus, however, is revealed when we apply it to rates of change of entities with respect to quantities that may or may not be time; this extends the rate of change concept considerably and enlarges the application of the differential calculus enormously. As we have already stated, this extension of the calculus to any kind of rate of change of a function of another quantity was developed by Leibniz. We present a few simple examples of this concept which is important to us as car drivers. We spoke above of the time rate of the burning of gasoline in our cars, but we generally think of that as a mileage rate rather than a time rate and speak of the number of miles per gallon, or, equivalently, the number of gallons per mile. Thus, we may speak of 20 miles per gallon or one-twentieth of a gallon per mile. As another example of a rate of change which does not involve time explicitly, we consider a planet moving in an ellipse around the sun. We can give its position at any moment in polar coordinates by giving its distance $r$ from the sun at any moment and the angle $\theta$ that the radius vector $r$ makes with the major axis (lying along the $x$ axis) of the ellipse. We saw in the last chapter that $r$ and $\theta$ (actually the cosine of $\theta$) are related by a simple equation; $r$ is thus a function of $\theta$, as defined by that equation, so that it depends on $\theta$. What is the rate of change of $r$ with respect to $\theta$? In other words, what is the value of $\lim_{\Delta\theta \to 0} \Delta r/\Delta\theta$, or, as Leibniz expressed it, the value of $dr/d\theta$? The calculus shows us how to find this limit (the derivative of $r$ with respect to $\theta$) from the equation that connects $r$ and $\theta$ (the functional relationship between the two entities).

Later we show that such instantaneous rates of changes or derivatives of functions with respect to the variables they may depend on can be deduced graphically, but here we investigate these derivatives (the basis of differential calculus) analytically. Once we understand derivatives, working with the differential calculus becomes routine. We do this by returning to our rectangular Cartesian coordinate system $x$, $y$ and considering any continuous function $y = f(x)$ which we can plot in the $x$-$y$ plane in accordance with the nature of the function $f(x)$ which tells us the value that $y$ has for any given value of $x$. Finding the rate of change of $y$ with respect to $x$ at any point $x$ involves first comparing the $y$ value at that point with the $y$ value at a point very close to $x$ (either a bit smaller or a bit larger than $x$) and then seeing how the difference between these two $y$ values compares with the difference between the two $x$ values. We now express this first step arithmetically as follows: At the point $x$, which may be any point on the $x$ axis, the value of $y$ is $f(x)$, and at the nearby point $x + \Delta x$, where $\Delta x$ is a very small increase (or decrease) in $x$ (also called an increment in $x$), the value of $y$ is $f(x + \Delta x)$. This means that to obtain the value of $y$ at the nearby point $x + \Delta x$, we must replace $x$ by $x + \Delta x$, wherever $x$ appears in the function $f(x)$. If the function $f(x)$ is $ax + bx^2$, as a simple example, $f(x + \Delta x)$ is $a(x + \Delta x) + b(x + \Delta x)^2$. We now compare these two values of $y$ by subtracting one from the other: $f(x + \Delta x) - f(x)$. If we now divide this difference by the difference $\Delta x$ between the two $x$ values of the points, we obtain $[f(x + \Delta x) - f(x)]/\Delta x$, which tells us the average rate of change of $y$ [or $f(x)$] taken over a small $x$ interval $\Delta x$, but this is still not the instantaneous rate of change of $y$ with respect to $x$ or the derivative of $y$ or $f(x)$ with respect to $x$.

To obtain that important mathematical entity we must allow the nearby point to approach $x$ until it is at an infinitesimal distance $dx$ from $x$; we express this relationship by writing $\lim_{\Delta x \to 0} [f(x + \Delta x) - f(x)]/\Delta x$. The meaning of this expression is quite clear. First, we must replace $x$, wherever it appears in $f(x)$, by $(x + \Delta x)$; we then subtract $f(x)$ from it, and divide this difference by $\Delta x$ and allow $\Delta x$ to go to zero. The difference $f(x + \Delta x) - f(x)$ then becomes the differential $df(x)$ (or $dy$) and $\Delta x$ becomes $dx$ so that we obtain, in the limit $\Delta x \to 0$, the derivative of the function $f(x)$ with respect to $x$ written as $df(x)/dx$ or $dy/dx$. The arithmetic or algebraic operations we have just outlined are quite straightforward and present no conceptual difficulty but they leave us somewhat uneasy because it appears that some kind of trick or intellectual sleight of hand is being performed since just placing $\Delta x = 0$ in the fraction $[f(x + \Delta x) - f(x)]/\Delta x = [f(x) - f(x)]/0 = 0/0$ and is therefore meaningless. But this difficulty arises because the final operation of allowing $\Delta x$ to become zero must be carried out only after the fraction $[f(x + \Delta x) - f(x)]/\Delta x$ has been reduced to its simplest form by canceling $\Delta x$ in the denominator against any factor $\Delta x$ in the numerator. When this is done, the denominator in the final expression is 1 so that it cannot equal zero when $\Delta x$ goes to zero. Thus, the day is saved.

To illustrate this point we consider the simple example $y = f(x) = x^2$. Following the algebraic directions given above we have $f(x + \Delta x) = (x + \Delta x)^2$ so that

$$f(x + \Delta x) - f(x) = (x + \Delta x)^2 - x^2$$

$$= [x^2 + (2x)(\Delta x) + (\Delta x)^2] - x^2$$

and

$$\frac{f(x + \Delta x) - f(x)}{\Delta x} = \frac{[x^2 + (2x)(\Delta x) + (\Delta x)^2] - x^2}{\Delta x}$$

$$= \frac{(2x)(\Delta x) + (\Delta x)^2}{\Delta x}$$

If we place $\Delta x = 0$ in this fraction we obtain $[(2x)(0) + (0)^2]/0 = 0/0$ and therefore, a meaningless expression, but we must not do this until after we have canceled out all factors $\Delta x$ in the numerator—which we can write as $(\Delta x)(2x + \Delta x)$. The fraction thus becomes $(\Delta x)(2x + \Delta x)/\Delta x$. After canceling $\Delta x$ in the numerator and denominator, we obtain $(2x$

$+ \Delta x$). We may now place $\Delta x = 0$ without doing any mathematical damage and we obtain $2x$. We thus see that $dy/dx = df(x)/dx = d(x^2)/dx = 2x$, which is the rate of change of $x^2$ or the derivative of $x^2$ with respect to $x$ at the point $x$. Note that in all this algebraic maneuvering, $\Delta x$ is treated as an ordinary algebraic entity as are $dx$ and $dy$; we may add them, subtract them, multiply them, divide them, and so on.

The procedure described above for obtaining rates of change or derivatives of functions of variables is the rigorous mathematical technique which always gives the derivative of the function if the derivative exists. But we now approach this problem from the graphical point of view by studying the curve we get in the $x$-$y$ plane when we plot the function $y = f(x)$ against the variable $x$. Each point on this curve has the ordinate $f(x)$ and the abscissa $x$; we obtain the ordinate $y$ of a point on the curve by simply plugging the numerical value of its abscissa $x$ into the function $f(x)$, a purely arithmetic operation. We now consider any point on the curve with coordinates $x$, $y$ and a neighboring point with coordinates $(x + \Delta x)$, $(y + \Delta y)$; the neighboring point is $\Delta x$ units of length to the right of the first point and $\Delta y$ units of length above the first point, where $\Delta x$ and $\Delta y$ are very small but finite quantities. Here we consider $f(x)$ and $x$ as distances but they need not be distances; $f(x)$ may be any kind of entity we please (for example, the price of certain goods) and $x$ may be some other entity on which $f(x)$ (the price) is dependent (such as the demand for such goods). The points on the $y$ and $x$ axes then represent money so that the number 5 on the $y$ axis means a price 5 times higher than the number 1 on the $y$ axis and the number 3 on the $x$ axis means a demand (in terms of money) 3 times greater than that meant by the number 1 on the $x$ axis. With this understood, we return to our curve in the $x$-$y$ plane and describe the coordinates of points on the curve as distances, keeping in mind that these may be metaphors (scale factors) for quite different quantities. Here we see the power of mathematics which can use theorems in geometry to describe relationships among quantities that have nothing to do with geometry. Considering the two neighboring points $(x, y)$ and $(x + \Delta x, y + \Delta y)$ on our $f(x)$ curve, we recall that we began our study of the rate of change of $f(x)$ (or $y$) with respect to $x$ by noting that the ratio $\Delta y/\Delta x$ for the two neighboring points is the average rate of change of $f(x)$ taken over the small stretch of the curve that goes from the point $x, y$ to the neighboring point $(x + \Delta x)$, $(y + \Delta y)$. If instead of going along the curve from the starting point on the curve to the neighboring point, we move along the straight line connecting these two points, we can interpret the ratio $\Delta y/\Delta x$ trigonometrically. The line connecting the two points is the hypotenuse of a very tiny right triangle whose vertical side (parallel to the $y$ axis) is $\Delta y$, the difference between the $y$ values (ordinates) of the two points, and whose horizontal side (parallel to the $x$ axis) is $\Delta x$, the difference between the $x$ values (abscissas) of the two points. The ratio $\Delta y/\Delta x$ is therefore related to the sine and cosine of the angle which the straight line (the hypotenuse of the small right triangle) makes with $\Delta x$ or with the $x$ axis if that straight line is extended to cut the $x$ axis. The sine of the angle is $\Delta y/$hypotenuse and the cosine of this angle is $\Delta x/$hypotenuse so that $(\Delta y/$hypotenuse$)/(\Delta x/$hypotenuse$) = \Delta y/\Delta x =$ sine of angle/cosine of angle $=$ tan of angle.

We can now see exactly what the derivative means and how it is to be calculated graphically. We recall that the derivative $dy/dx = \lim_{\Delta x \to 0} \Delta y/\Delta x$, which means that we must allow the neighboring point $x + \Delta x, y + \Delta y$ to approach the point $x, y$ to obtain the derivative, but this means that the straight line connecting the two points coincides ever more closely with the stretch of the curve connecting the two points as we perform this operation. At the same time, as we push the neighboring point closer to the initial point $x$, $y$, the angle which the straight line makes with the $x$ axis changes gradually until the two points coincide and the ratio $\Delta y / \Delta x$ becomes the derivative $dy/dx$; the straight line then touches the curve only at the initial point $x$, $y$, and at no other point, becoming what is called the tangent of the curve at that point. If the angle between this tangent and the $x$ axis is $\theta$ at the point where the tangent cuts the $x$ axis, then the derivative of the function with respect to $x$ is $dy/dx = df(x)/dx = \sin \theta / \cos \theta = \tan \theta$. This number is called the slope of the curve $y = f(x)$ at the point $x$. It gives the rate at which $y$ is changing with respect to $x$ at that point.

We now have a simple way of obtaining the derivative of any function of the variable $x$ with respect to $x$ [the rate of change of the function $y(x) = f(x)$ with respect to $x$]: we plot the function in the $x$-$y$ coordinate system to obtain a curve of the function in the $x$, $y$ plane, and then we draw the tangent to any point on that curve. We then measure the angle that tangent makes with the $x$ axis and look up its tangent in a trigonometric table of sines, cosines, and tangents. This number is the derivative of the function with respect to $x$ at the point $(x, y)$. We now recognize the derivative as the slope of the curve at the point being considered. Depending on the mathematical form or shape of the curve, the slope may change from point to point so that the slope itself changes from point to point with respect to $x$, which means that the function exhibits a second rate of change (the rate of change of the rate of change). We discuss higher rates of change of a given function of $x$ further on in this chapter.

Defining the rate of change of a function (its derivative) in terms of the curve's slope at any point reveals some important applications of the derivative to the solution of problems in mathematics, physics, chemistry, and astronomy. These are known as maxima and minima problems. In the chapter on algebra we pictured the plot of a polynomial in the $x$-$y$, plane [the curve $y = f(x)$ where $f(x)$ is the polynomial] as a kind of road which we follow along the horizontal $x$ axis from west to east starting from the $y$ axis. We noted there that the slope of the road changes constantly as it rises and falls with respect to the horizontal, which we know to be true from our car-driving experiences. A road that is rising at one point of our trip must fall at another point unless it leaves the earth entirely—which is a ridiculous notion. This means that at some point between rising and falling the road is at a maximum height above the horizontal and the road must then begin to fall. By the same kind of reasoning we see that the road cannot continue falling indefinitely but must begin to rise. Therefore, at some point on the road between falling and rising the road is at a minimum height above the horizontal (it may be at the horizon or even below it).

We now apply these road observations to the function $y = f(x)$ which the road represents, and we see at once that we are talking about the derivative $f(x)$ with respect to $x$. If the road is rising at any point $x$, its slope at that point, given by $dy/dx = df(x)/dx$, is positive; the derivative, with respect to $x$, of the function (the tangent of the angle that the tangent line at the point makes with the $x$ axis) is positive. Where the road is falling, the derivative (the slope) is negative. Where the road is neither rising nor falling, it must be parallel to the $x$ axis (the horizontal); its slope at that point must be zero and its tangent line at that point must be parallel to the $x$ axis. If we have the functional relationship $y = f(x)$ between two entities $y$ and $x$, we can see at once what values of $x$ give maximum or minimum values of $y$ by plotting $f(x)$

in the $x$-$y$ plane and seeing where the hills and valleys of the function are located or we can differentiate the function with respect to $x$ (find the derivative in accordance with the procedure described above) and set this derivative equal to zero and solve it for its various roots as described in the algebra chapter.

We now apply these ideas to specific functions to illustrate the simplicity of the calculus. We consider first the function $y = a$ where $a$ is a fixed number. The function is thus a constant, independent of $x$; if we plot it in the $x$-$y$ plane, we obtain a straight line which is always at the same distance $a$ from the $x$ axis. Considering it as a road, we see that it is a horizontal road on a plateau at a height $a$ above the zero height horizontal. Since this road has no hills or valleys, its slope is always zero, which means that $dy/dx = 0$. If the derivative of a function is zero for all values of $x$, the function is a constant and independent of $x$.

We consider now the function $y = x$. We see that if we increase $x$ by a small amount $\Delta x$, then $y$ increases by exactly the same amount so that $\Delta y = \Delta x$ and $\Delta y/\Delta x = 1$ and this is true regardless of the smallness of $\Delta x$ so that for $\Delta x = 0$ we have $dy/dx = 1$ since $dy = dx$. The slope of the function everywhere (for every value of $x$) equals 1; an advance of a given amount along the $x$ axis equals the rise along the $y$ axis, which is obvious from the graph of $x = y$: a straight line through the origin at a 45° angle with respect to the $x$ axis.

We consider now the function $y = ax$, where $a$ is a constant (a definite number), whose plot like that of $y = x$ is a straight line through the origin 0 but whose slope is $a$—not 1. For a change $dx$ in $x$ we have a change $dy$ in $y$ such that $y + dy = a(x + dx)$, or $dy = a(dx)$, since $y = ax$ so that $y$ and $x$ cancel each other. Dividing both sides by $dx$ we obtain $dy/dx = a$. If we multiply a function of $x$ by a constant (a quantity that does not depend on $x$), the derivative of the product of the constant and the function is the derivative of the function multiplied by the constant: $(d/dx)[af(x)] = a\,df(x)/dx$. In differentiating we leave constant factors alone. A constant factor merely multiplies the slope of a function at each point $x$ by that factor. Thus, slope (factor times function) equals factor times slope of function.

Instead of speaking of the slope $a$ of the straight line $y = ax$, we can speak of the pitch of the line, as discussed in some detail in our chapter on algebra, from an algebraic point of view. Although we are all acquainted with pitch, we do not ordinarily associate it with calculus but a little thought reveals that it is the essence of calculus. If we plot any function $y = f(x)$ in the $x$-$y$ plane and isolate a tiny piece of the plotted curve in the neighborhood of any $x$ value, that tiny piece is very nearly a straight line that has a certain pitch. If we take an infinitesimal piece $ds$ of the curve (the use of $s$ means that we measure the length of the tiny piece along the curve), this is essentially a piece of a straight line whose pitch (slope) equals $dy/dx$ at that point, which, of course, is the derivative of the function $f(x)$ at the point $x$.

Since $ds$ is essentially an infinitesimal straight line, it is the hypotenuse of an infinitesimal right triangle of which the infinitesimal $dy$ and $dx$ are the other two sides. Applying the Pythagorean theorem to this infinitesimal triangle we obtain the differential equation $(ds)^2 = (dx)^2 + (dy)^2$ (an equation of differentials or infinitesimals). We note that even though $ds$, $dx$, and $dy$ are infinitesimals, they obey the ordinary rules of algebra. By rearranging the terms on the right-hand side somewhat, we obtain the equation

$$(ds)^2 = (dx)^2[1 + (dy/dx)^2]$$

and, on taking the square root of both sides, we obtain

$$ds = dx\sqrt{1 + (dy/dx)^2}$$

This is a very interesting and important equation because it tells us that if we know the slope $(dy/dx)$ of a function $y = f(x)$ at each point $x$, we can reconstruct the curve of the function plotted in the $x$-$y$ plane. We simply square the slope, add 1 to it, and take the square root of this sum. We now multiply this square root by the infinitesimal $dx$ along the $x$ axis. This product is the length $ds$ of the infinitesimal piece of the curve in the neighborhood of the point $x, y$ on the curve. Since $dy/dx$ is the given slope of this tiny piece of the curve, we can lay the piece off in the right direction, as dictated by the slope. By doing this for each of the values of $x$ for which $dy/dx$ is given, we can lay off all the infinitesimal pieces, one after the other, and thus construct the curve that is the plot of the function $y = f(x)$. This procedure, the inverse of differentiation, called integration, is treated as integral calculus in the second half of the typical calculus course; we discuss it later in this chapter. Our discussion here of the properties of the curve of a known function shows us that calculus is essentially the algebra and geometry of infinitesimal quantities.

Returning now to the pitch concept we note its many applications to familiar things such as a roof (the pitch of a roof) and its incorporation into structures and tools such as a propeller, the threads of a screw, a spiral staircase, the blades of a fan, etc. In all these examples, the basic principle is the same as that for the straight line $y = ax$.

We now consider the function $y = ax + b$ where $b$ is a constant just like $a$; $b$ does not depend on $x$ and so it does not change when $x$ is altered. The plot (or graph) of this function is a straight line parallel to the line $y = ax$, but shifted upward or downward a distance $b$, so that every point on this line is shifted vertically by an amount $b$ from the equivalent point (the two points have the same $x$ coordinate) on the line $y = ax$. If we now change $x$ by an amount $dx$ in the function $y = ax + b$, the amount $dy$ by which $y$ changes is simply $dy = adx$ since $b$ suffers no change, no matter how we change $x$. If we now divide both sides of the equation $dy = adx$ by $dx$, we obtain $dy/dx = a$ so that the derivative of the function $f(x) = ax + b$ with respect to $x$ is the same as that of the function $f(x) = ax$; both functions have the same slopes at equivalent points. This tells us that adding a constant to a function of any variable does not alter the derivative of that function with respect to the variable. Put differently, the slope of a function at any point on the curve of the plot of the function is not altered by the addition of a constant.

Having exhausted the calculus of the simple function we proceed to the function $y = f(x) = ax^2$, which is quadratic in the variable $x$, with $a$ as a constant. As we saw in our algebra chapter, this is the equation of a parabola that just touches the $x$ axis at the origin $y = x = 0$ and whose two branches move off symmetrically to infinity, as $x$ goes off to $\pm$infinity to the right and left of the $y$ axis, respectively. Because the power of the variable $x$ in the equation is larger than 1, the rate of change of $f(x)$ with respect to $x$ [the derivative $df(x)/dx$] varies from point to point on the parabola. We can deduce this derivative by the standard algebraic procedure by calculating the expression $\lim_{\Delta x \to 0} [f(x + \Delta x) - f(x)]/\Delta x$ which is fairly easy to do. However, deducing the derivative by reasoning and using some basic principles is more enjoyable and instructive than performing the calculations themselves.

Since $y$ and $x$ in the equation $y = ax^2$ may be any quantities we please without changing the derivative $dy/dx$, we choose them to be distances. We then see that the constant $a$ must be the reciprocal of a distance because $x^2$ is the square of a distance (an area) whereas $y$ is a distance. By making $a$ the reciprocal of a distance (1/distance), we make the product $ax^2$ become distance$^2$/distance (which equals distance) so that the left-hand side $y$ (a distance) of our equation and the right-hand side $ax^2$, now also a distance, are dimensionally the same (they have the same spatial attributes). We can now write down $dy/dx = (d/dx)(ax^2)$ by inspection. Since the expression $dy/dx$ is a distance divided by a distance ($dy$ and $dx$ are both distances even though they are infinitesimals), the left-hand side of the differential equation has no spatial dimensions (they cancel out); hence, the right-hand side of the equation must also be dimensionless which means that the derivative $(d/dx)(ax^2)$ can only contain $a$ and $x$ as the product $ax$, because that is the only dimensionless combination of the constant $a$ and the variable $x$ we can construct. But we can still multiply the quantity $ax$ by a pure (non-dimensional) number without destroying the dimensional harmony of the two sides of the equation.

To find this numerical coefficient we note that $(d/dx)(ax^2)$ equals $a(d/dx)(x^2)$, since $a$, a constant, does not change; we leave it alone as an unchanging factor in calculating the rate of change of $ax^2$ with respect to $x$. To calculate the rate of change $(d/dx)(x^2)$ of $x^2$ with respect to $x$ we write $x^2$ as $(x)(x)$ so that $d(x^2)/dx = (d/dx)[(x)(x)] = (d/dx)(xx)$. We note that both factors $x$ in the product $xx$ change when $x$ changes, but we can consider this change as occurring in two steps: the first factor $x$ changes while the second one remains constant, and then the second factor changes while the first one remains constant. The correct rate of change of the product $xx$ or $x^2$ is then the sum of these two separate changes. We thus have

$$(d/dx)(x^2) = (d/dx)(xx) = (dx/dx)(x) + (x)\,dx/dx$$

But $dx/dx = 1$, so that we have $(d/dx)(x^2) = (1)(x) + (x)(1) = 2x$. The numerical coefficient we seek is thus 2. If we apply this product rule to the rate of change of $x^3 = xxx = x^2x$ with respect to $x$ we have

$$(d/dx)(x^3) = (d/dx)(x^2x) = (dx^2/dx)x + x^2(dx/dx)$$

But $dx^2/dx = 2x$ and $dx/dx = 1$ so that we obtain

$$(d/dx)(x^3) = (2x)x + x^2(1) = 2x^2 + x^2 = 3x^2$$

We can proceed in the same way with $(d/dx)(x^4)$ by writing $x^4 = x^3x$ and then applying our rule for the rate of change of a product so that

$$(d/dx)(x^4) = (d/dx)(xx^3) = (dx/dx)x^3 + x(dx^3/dx)$$
$$= x^3 + (x)(3x^2) = x^3 + 3x^3 = 4x^3$$

From these equations we see a general rule emerging for the rate of change of $x$ raised to any power. We saw that the rates of change of $x$ raised to the powers 1, 2, 3, 4 ($x$, $x^2$, $x^3$, $x^4$) are 1, $2x$, $3x^2$, $4x^3$, respectively. We therefore surmise correctly that the rates of change (with respect to $x$) of $x^5$, $x^6$, $x^7$ are $5x^4$, $6x^5$, $7x^6$, and so on. The rate of change with respect to $x$ for any power of $x$ such as $x^n$ (where $n$ may be any negative or positive integer, fraction, irrational, real, complex, or imaginary number) is found by reducing the exponent of $x$, $n$, by 1 and multiplying $x$ to the reduced power (exponent) $n$. This rule may be written algebraically as $(d/dx)(x^n) = nx^{n-1}$.

The importance and usefulness of this rule cannot be overstated; it enables us to find the rate of change $(d/dx)f(x)$ of any function of $x$ with respect to $x$ because every such function is a polynomial in $x$ (a finite sum of powers of $x$ multiplied by appropriate constant coefficients) or an infinite power series in $x$ (a sum of an infinite number of powers of $x$ starting with $x^0$, with each succeeding power increased by 1). We illustrate the usefulness of this power (or exponent) rule with a few simple examples, taking as the first example $y = (1/x)^n$. We seek the formula for

$$dy/dx = (d/dx)(1/x)^n = (d/dx)(1/x^n)$$

But $1/x^n = x^{-n}$ so that $(d/dx)(1/x^n) = (d/dx)(x^{-n})$. Applying our power (exponent) rate of change rule we then have

$$(d/dx)(1/x)^n = (d/dx)(x^{-n}) = -nx^{-n-1}$$
$$= -nx^{-(n+1)} = -n/x^{n+1}$$

We can prove this equation very easily using the product rule, because we have $y = 1/x^n$, and this equation becomes $x^n y = 1$ on multiplying both sides by $x^n$, according to the rules of algebra. We now take the derivatives (rates of change) of both sides of this equation, noting that the right-hand side is just a numerical constant so that its rate of change is zero: $(d/dx)(x^n y) = 0$. Applying the product rule to the left-hand side rate of change we obtain the equation $(d/dx)(x^n)y + x^n(dy/dx) = 0$ or $x^n(dy/dx) = -y(dx^n/dx)$. Since $y = 1/x^n$ we replace $y$ by $1/x^n$ on the right and obtain $x^n(dy/dx) = (-1/x^n)(dx^n/dx) = (-1/x^n)(nx^{n-1}) = -nx^{n-1}/x^n = -n/x$. If we divide both sides by $x^n$, we obtain $dy/dx = -n/x^{n+1}$, which proves that $(d/dx)(1/x^n) = -n/x^{n+1}$.

The second instructive example of using the product rule to obtain the rate of change with respect to $x$ of a function of $x$, where the function is such that applying the standard procedure leads to algebraic complications, is finding the rate of change of $y = \sqrt{(1 + x)}$. For the rate of change of $y$ we have $dy/dx = (d/dx)\sqrt{(1 - x)}$. We know that the rate of change of 1 is zero and the rate of change of $x$ is 1, but that does not do us much good because we cannot apply the rate of change procedure directly to the expression $(1 + x)$ inside the square root. We avoid that difficulty by multiplying both sides of the equation $y = \sqrt{(1 + x)}$ by $\sqrt{(1 + x)}$ to obtain $\sqrt{(1 + x)}y = \sqrt{(1 + x)}\sqrt{(1 + x)}$. We now equate the derivatives of both sides of this equation to obtain $(d/dx)[\sqrt{(1 + x)}]y = (d/dx)(1 + x)$. On applying the product rule to the left-hand side we obtain

$$(d/dx)[\sqrt{(1 + x)}]y + \sqrt{(1 + x)}\, dy/dx = dx/dx = 1$$

But $y = \sqrt{(1 + x)}$ and $dy/dx = (d/dx)\sqrt{(1 + x)}$ so that we obtain from the left-hand side of the previous equation

$$[(d/dx)\sqrt{(x + 1)}]y + (dy/dx)\sqrt{(1 + x)}$$
$$= 2\sqrt{(1 + x)}[d\sqrt{(1 + x)}/dx]$$

But this left-hand side equals the right-hand side 1. Hence, $2\sqrt{(1 + x)}d\sqrt{(1 + x)}/dx = 1$. Dividing both sides by $2\sqrt{(1 + x)}$ we obtain $d\sqrt{(1 + x)}/dx = 1/2\sqrt{(1 + x)}$. These examples teach us that regardless of the complexity of the function $y = f(x)$, we can always obtain a manageable algebraic expression by permissible algebraic manipulations for which the derivative or rate of change with respect to $x$ can be written down.

Thus far we have been discussing the rates of change of algebraic functions of $x$ with respect to $x$, but what about the rates of change of trigonometric functions of a variable with respect to that variable? For example, what is the rate of change of $\cos \theta$ or $\sin \theta$ with respect to $\theta$, where $\theta$, a pure number, is expressed in radians? We can obtain the rates of change of cos $\theta$ and sin $\theta$ with respect to $\theta$ by going back to our unit circle and the definitions of sin $x$ and cos $x$ in terms of the geometric properties of that circle. Taking the origin 0 of our Cartesian coordinates as the center of the unit circle, we defined the $y$ coordinates of a point P on the circumference of the unit circle as the sine of the angle $\theta$ that the radius of the circle from the center 0 to P makes with the $x$ axis, and the $x$ coordinate of that point P as the cosine of $\theta$.

If we now increase $\theta$ by an infinitesimal amount $d\theta$, we obtain a point P' on the circumference which is infinitesimally close to P and whose $x$ coordinate is smaller than that of P by the amount $dx$ and whose $y$ coordinate is larger than that of P by the amount $dy$. The arc $d\theta$ from P to P' is the hypotenuse of an infinitesimal right triangle of which $dx$ and $dy$ are the two other infinitesimal sides (at right angles to each other); one of the angles in this infinitesimal triangle is $\theta$ and the other is $(90° - \theta)$. Since the infinitesimal side $dx$ is opposite the angle $\theta$ in this triangle and the infinitesimal side $dy$ is adjacent to $\theta$, the sine of $\theta$ (opposite over hypotenuse) and its cosine (adjacent over hypotenuse) equal $dx/d\theta$ and $dy/d\theta$, respectively. But we must ask whether these rates of change of $y$ and $x$ with respect to $\theta$ are positive or negative. If $y$ increases as $\theta$ increases (if the ordinate of P increases as P goes to a point farther from near the $x$ axis to one near the $y$ axis), then $dy/d\theta$ must be positive. Since this is so for $y$ and $\theta$ (they increase together), $dy/d\theta$ is positive; hence $dy/d\theta = +\cos \theta$. But $x$ decreases as $\theta$ increases; hence $dx/d\theta$ is negative so that $dx/d\theta = -\sin \theta$. From our definitions of the sine and cosine of the angle $\theta$ in our unit circle as the coordinates $y$ and $x$ respectively of P, we see that $d \sin \theta/d\theta = dy/d\theta = \cos \theta$ and $d \cos \theta/d\theta = dx/d\theta = -\sin \theta$. Thus, the derivative of the sine of an angle with respect to the angle (its rate of change with respect to that angle) equals the cosine of the angle, and the derivative of the cosine with respect to the angle equals the negative of the sine of the angle.

We complete our discussion of the differentiation of functions with respect to variables (derivatives) by writing down, without proof, the derivatives of the logarithm of $x$ and the exponential $a^x$ with respect to $x$, where $a$ is a constant. The rate of change of the logarithm of a quantity $x$ as $x$ changes equals $1/x$; if we plot the function $y = f(x) = \log x$, we obtain a curve whose slope at each point on the curve equals the reciprocal of the $x$ coordinate at that point. The derivative of the function $y = f(x) = a^x$ with respect to $x$ is a constant times $a^x$; if we plot $y = a^x$, the slope at each point on the plotted curve equals the $y$ coordinate of that point multiplied by the constant.

In our discussion of the derivative (the rate of change of one quantity with respect to another on which the first quantity depends), we emphasized the importance of keeping in mind that the concept of the derivative has meaning only if we specify it as a rate of change with respect to some particular entity. This entity is usually specified in the functional relationship. If our function is written as $y = f(x)$, it represents an explicit relationship between $y$ and $x$. The derivative of $y$ with respect to $x$ is automatically assumed if we speak about the rate of change of $y$. But an implicit rate of change of $y$ with respect to another variable may exist. If $x$ itself depends on the time, $y$ also depends on the time through $x$ by a kind of functional chain which attaches $y$ to $x$ and $x$ to $t$ ($t$ stands for time) through $x$. A rule tells us how to obtain the rate of change of $y$ [which may be written as $f(x)$] with respect to $t$ if we know how $x$ depends on $t$; this rule, appropriately enough, is called the "chain rule." The rule is simply an algebraic relationship among the infinitesimals $dy$ $[=df(x)]$, $dx$ and $dt$; it is a product of two fractions constructed from these infinitesimals—the rate of change of $f(x)$ with respect to $x$ and the rate of change of $x$ with respect to $t$: $df(x)/dx$ and $dx/dt$. By multiplying these expressions together, we obtain the product $[df(x)/dx]$ $(dx/dt)$. Since $dx$ in the denominator of the first fraction cancels $dx$ in the numerator of the second fraction, this product exactly equals $df(x)/dt$, which is exactly the expression we want. If we move along a road at a certain speed $v = dx/dt$, we note the slope of the road at each distance $x$ from our starting point or at each moment of our journey. We obtain the rate at which the slope changes in time for us by multiplying the rate at which the slope changes with distance by our speed $v$ $(dx/dt)$.

Although the differential calculus is a powerful analytical tool which enables us to do mathematics with infinitesimals, it is a very simple mathematical technology. As we have stated previously, it is an analytical technique for finding the slope at any point of a curve which is a graphical representation of the relationship of one quantity to another. If this curve can be plotted very accurately, we obtain the slope (the rate of change of the curve at that point) by drawing a straight line that cuts the $x$ axis and touches the curve at just this point. The tangent of the angle that this line (the tangent to the curve at the point) makes with the $x$ axis is the slope of the curve at the point and, therefore, the value of the derivative at that point of the function $y = f(x)$ represented by the curve. It is the value of the mathematical entity $df(x)/dx$ when $x$ in this expression is placed equal to the $x$ coordinate of the point. For example, if $f(x) = 2x^2$, then $df(x)/dx$ $= (2)(2x) = 4x$, and if the point we are talking about has the coordinates $x = 2$ and $y = f(2) = 8$ [we replace $x$ by 2 in $f(x)$ to obtain $y$], the derivative $dy/dx = 4x$ has the value 8 (4 times 2) at this point. The tangent to the curve at this point cuts the $x$ axis at an angle whose tangent is 8 (about 83°). This is the essence of differential calculus; all other aspects and features of calculus can be deduced from this statement.

If we follow a curve continuously along the $x$ axis, we note that its slope, in general, changes continuously so that the derivative $dy/dx$ changes continuously. We may therefore speak of the rate of change with respect to $x$ of $dy/dx$. Treating this derivative as a function of $x$ may permit us to define mathematically its rate of change with respect to $x$ as $d/dx[df(x)/dx] = d[df(x)]/(dx)^2$. The two $d$'s in the numerator are combined and written as $d^2$ so that the numerator of this fraction is written as $d^2f(x)$ and the rate of change of the rate of change is written as $d^2f(x)/(dx)^2$ or $d^2f(x)/dx^2$. This is called the second derivative of the function $f(x)$ with respect to $x$. The numerator in this expression is not a square of anything; it is the differential of a differential. The denominator, on the other hand, is a real square of an entity—the square of $dx$ even though it is written as $dx^2$. These points about the nomenclature of calculus must be kept in mind in reading the literature.

Just as we can have the second rate of change with respect to $x$, we can have still higher rates of change (third, fourth, and so on) which we write as $d^3f(x)/dx^3$, $d^4f(x)/dx^4$, etc. We represent all such rates of change with respect to $x$ by the expression $d^nx/dx^n$, where $n$ may be any positive integer we please; fractions and negative integers are excluded.

We come now to the second branch of the calculus, the integral calculus, which has a far greater range of application in science than the differential calculus. Before we discuss the importance of the integral calculus to science, we consider it from a purely mathematical point of view and introduce integration and the integral as a mathematical operation and a mathematical entity, respectively. Just as the differential calculus is based on finding the derivative of a function with respect to the variable it depends on by differentiating the function with respect to that variable, so the integral calculus is based on finding the integral of a function by integrating it over a given range of the variable it depends on.

We can best illustrate the integration concept or the integral by studying how they are related to or stem from differentiation and the derivative of a function. We therefore consider a function $f(x)$ of $x$, the values of whose derivative we know for all values of $x$ from 0 to some unspecified value $x$ ($x$ may be any number we please). We do not know the function $f(x)$ but we know its derivative $df(x)/dx$ with respect to $x$. The integral calculus answers or tries to answer the question "What is $f(x)$?" This question can be answered analytically, in the sense of finding a definite mathematical expression for $f(x)$ in terms of $x$, only rarely. We can certainly set up any array of slopes (derivatives for the different values of $x$) we choose and connect these by a curve (whose slopes at its various points equal the given slopes) but that does not mean that we can express this curve by a known function $f(x)$. In other words, the integral $f(x)$ may not exist as a closed mathematical expression. We see that the mathematical road $f(x) \rightarrow df(x)/dx$ is generally a one-way road. Given a mathematical function $f(x)$, we can (with very few exceptions) find the derivative $df(x)/dx$, but we cannot, in general, go from the derivative back to the function of which $df(x)/dx$ is the derivative. If we can do this, we see that, mathematically speaking, the integral is the inverse of the derivative since it is finding the function $f(x)$ which when differentiated is the given derivative.

To see what these statements mean, we consider the equation $df(x)/dx = F(x)$, where $F(x)$, a given function of $x$, has a definite value for any value of $x$ in a given range of $x$ values. For example, $F(x) = ax^2$ might be such a function where $a$ is a constant. Leaving the exact form of $F(x)$

unspecified, we consider the equation $df(x)/dx = F(x)$ further and multiply both sides by $dx$ to obtain $(dx)[df(x)/dx] = df(x) = F(x)\,dx$. The infinitesimal $df(x)$ or $F(x)\,dx$ is an infinitesimal piece of the function $f(x)$ we are seeking. We then obtain the function $f(x)$ by adding all these pieces together. The integration process is thus a sum and so we may show this explicitly by the equation $f(x) =$ sum of all $df(x)$'s $=$ sum $F(x)dx$. Instead of putting the word "sum" in front of $F(x)\,dx$, mathematicians place the symbol $\int$ (an enlarged s) in front, which is called the integral sign, and write $f(x) = \int F(x)\,dx$. If we keep in mind that this is just a sum of many (in principle, an infinite number of) infinitesimal pieces, we have no trouble understanding the concept of the integral.

We illustrate this summing operation (integration) by a simple example which also reveals another important aspect of the integral. We set up our rectangular Cartesian coordinate system $x$, $y$ taking $x$ and $y$ to be spatial distances so that the $x$-$y$ plane is an area. If we draw a curve of any kind in the plane from a point A to a point B, we define the "area under this curve" as the area bounded by this curve above, by the two perpendicular lines from the points A and B to the $x$ axis on the left and right, and by the stretch of the $x$ axis between these two perpendicular lines. We now consider the integral of the function $F(x) = x$ between the point 0 (the origin) and a point $x$, $x$ (the two coordinates of this point are both $x$) on the plane. The function $F(x) = x$ is the straight line through the origin we discussed previously; every point on it is equidistant from the $x$ and $y$ axes so that its two coordinates are equal. The integral $f(x) = \int_0^x F(x)\,dx = \int_0^x x\,dx$ is simply the sum of all the infinitesimal products $x\,dx$ from $x = 0$ to $x = x$; the 0 at the bottom of the integral sign $\int_0^x$ and the $x$ at the top of the integral sign define the stretch of the $x$ axis along which we integrate. We leave the $x$ at the top of the integral sign unspecified so that we may choose any value we please for it. The integral is thus a function $f(x)$ of $x$ which we seek. We now replace the integral by the sum $x_1 dx + x_2\, dx + \cdots + x_f\, dx$ which is its equivalent (and in fact its equal if an infinite number of terms is taken). This sum, for the integral we are now considering, $\int^x x\, dx$, is $0 dx + x_2\, dx + \cdots + x dx$ since $x_1$ is 0 and $x_f$ is $x$. We thus have $f(x) = \int_0^x x\, dx = 0 + x_2 dx + \cdots + x\, dx = (0 + x_2 + \cdots + x) dx$. Since the $x$ coordinates of the points on the straight line, which is the plot of our function, increase steadily by the same amount across each interval $dx$, we may replace each term in the parentheses by the average value of all of them and simply add up all these average values.

Since the first term of the sum is 0 and the last term is $x$, the average value is $x/2$ so that the integral equals $1/2(x + x + \cdots + x)\, dx$. We finally obtain the correct expression for the integral by equating the sum in the parentheses to the product of $x$ and the number of $x$'s in the parentheses. Since one $x$ must appear for each $dx$, this number is just the number of times $dx$ is contained in $x$—the quotient $x/dx$; the sum in the parentheses thus becomes $(x/dx)(x) = x^2/dx$. The integral $\int^x x\, dx$ therefore equals $x^2/2$. This result shows us two aspects of the integral $f(x)$. First, the integrand $F(x)$ of the integral equals the derivative with respect to $x$ of the integral itself

$$[F(x) = x = d/dx(x^2/2) = 1/2(d/dx)(x^2) = 1/2(2x) = x]$$

so that the integral is, indeed, the inverse of the derivative and vice versa. Second, the integral equals the "area under the curve" of the function $F(x)$ drawn from any definite point A on the $x$-$y$ plane to some indefinite point whose $x$ coordinate is $x$.

That the integral is, indeed, an area in the $x$-$y$ plane is exactly stated in its definition for it is a sum of products such as $F(x)\,dx$, where $dx$ is an infinitesimal distance and $F(x)$ is also a distance; namely, the $y$ coordinate of a point on the curve whose integral we seek. Thus, $F(x)\,dx$, as the product of two distances, is a tiny area and the integral is the sum of all these infinitesimal areas. In the example $\int_0^x x\,dx$, the total area defined is that of the equilateral right triangle, whose hypotenuse is the 45° straight line from 0 to the point $x, x$ and whose base and altitude both equal $x$; the area of this triangle is $1/2$ base times altitude $= (x/2)(x) = x^2/2$, which is the value of the integral.

We leave our discussion of the integral with one warning about the relationship between the integral of a function and the derivative of the integral as that function. The function $f(x) = \int_0^x x\,dx = x^2/2$ is not completely determined by the integral because the integral $\int_A^x x\,dx$ where A is a point on the straight line different from 0 also leads to the function $x^2/2$, but clearly the two integrals are not equal since they equal different areas. The functional dependence of the integral on $x$ is the same in both examples, but they differ by a constant area as we vary $x$: the constant area of the right triangle formed by the part of the line from 0 to the point A and the $x$ and $y$ coordinates of A. This simply means that the integral of any function $F(x)$ is determined only up to the extent of an additive constant. The two functions $f(x) = x^2/2 + \text{constant}$ and $f(x) = x^2/2$ have the same derivatives $x$ since the additive constant on the right-hand side drops out when we differentiate $f(x)$. To show that we must allow for an additive constant in integrating, we write the indefinite integral as $\int F(x)\,dx = f(x) + \text{constant}$.

If $F(x)$ is a sum of powers of $x$ (for example, the polynomial $a_0 + a_1x + a_2x^2 + \cdots + a_nx_n$), we can integrate each term separately and obtain the total integral as the sum of the integrals of the separate terms. The integral of any power of $x$ such as $a_n x_n$ is $[a_n/(n + 1)]x^{n+1}$; we can apply this rule to each term separately to obtain the final result. The integral of the polynomial above (except for an additive constant) is thus $a_0 x + (a_1/2)x^2 + \cdots + [a_n/(n + 1)]x^{n+1}$. The rule that the integral of any power $n$ (exponent) of $x$ equals $x$ raised to that power plus 1 (that is, $n + 1$), applies whether $n$ is negative or positive, an integer, a fraction, irrational, real, imaginary, or complex.

In introducing the integral calculus in this chapter, we stated that the integral calculus has a more extensive application to the solution of problems in science than does the differential calculus. To see why this is so, we devote the remaining pages of this chapter to the exploration of the relationship of mathematics to science, particularly to physics, which is the basic science from which all others branch out, and derive their basic laws. Physicists from the time of Kepler, Galileo, and Newton have pondered the mathematical description of the universe and wondered why the laws of nature can be expressed mathematically. Kepler, believing mathematics to be divine, stated that God played the mathematical game, that he "taught the game to nature," and "that all nature and the graceful sky are symbolized in the art of Geometria." Arithmetic and geometry came first, according to Kepler, because they were necessary for his satisfactory calculations of the planetary orbits, stating that "these figures pleased me because they are quantities, that is, something that existed before the skies. For quantities were created at the beginning, together with substance; but the sky was only created on the second day.... The ideas of quantities have been and are in God from eternity, they are God himself."

In recent years such great astrophysicists as Sir Arthur
Stanley Eddington and Sir James Jeans have suggested that
if a divine creator exists, he must be a mathematician. How
else, they asked, could a universe, governed by precise mathematical laws such as ours, have been created? Today such
clever mathematical physicists and cosmologists as Stephen
W. Hawking and John A. Wheeler are seeking a single
equation that describes both the structure and history of the
universe. But we need not go that far to accept the conclusion
that the laws of nature are most clearly expressed in mathematical forms or as mathematical relationships. Our universe
is a three-dimensional spatial manifold so that we can study
it only if we know the kind of geometry that governs it, which
for a long time was thought to be Euclidean (flat) geometry.
But geometric relationships can be understood in terms of
distances, surfaces, and volumes, which require arithmetic,
algebra, trigonometry, and calculus. We are thus thrust into
mathematics as soon as we begin to survey the universe. But
even before we consider the laws that govern the universe,
we must introduce a coordinate system to map events in space
and time whether we are studying cosmic events or laboratory
experiments so that analytic geometry is forced upon us. But
coordinate systems play a far greater role in science (physics)
than merely that of enabling us to plot relationships among
events. In the hands of the great theoretical physicists and
mathematicians, the coordinate system becomes the mathematical tool which can be used to gauge the truth or falsity
of hypotheses or theories.

As was first clearly enunciated and formulated into a
precise principle by Albert Einstein, natural laws are only
those statements about the universe that can be formulated
mathematically in such a way that the mathematical formulation remains the same in all coordinate systems. This principle

of invariance has played an enormous role in the development of modern physics. It is important because it is a kind of intellectual filter or sieve that separates correct statements about the universe (laws) from incorrect statements. To determine whether a statement is correct or not we need to verify whether or not it remains unaltered (invariant) when we shift from one coordinate system to another. But this seems like a hopeless task: How can we possibly carry out such a test, taking into account all kinds of possible coordinate systems? This difficulty, however, is more apparent than real for the number of generic coordinate systems is small and we can lump them together into a few groups.

To see the concepts and interrelationships that a coordinate transformation (shifting from one coordinate system to another) involves, we recall that we introduced the two-dimensional rectangular Cartesian coordinate system as consisting of two sets of mutually perpendicular straight lines and we labeled one set as $x$ lines and the other set as $y$ lines. Such a coordinate system implies that the geometry of the plane is Euclidean (flat) so that all spatial relationships are completely determined by or can be deduced from the Euclidean axioms. In particular, the square of the distance between any two points on the plane can be expressed in terms of the squares of the differences of the $x$, $y$ coordinates of those points, according to the Pythagorean theorem, which in turn stems from the Euclidean axioms. We can introduce a slew of other coordinate systems, each with its own two sets of intersecting lines which may or may not be at right angles to each other, which may or may not be straight (such as a polar coordinate system which contains a set of concentric circles and radiating straight lines), which may be oriented (rotated) in any direction whatsoever, and which may take any point on the plane as its origin. Regardless of which of these coordinate systems we adopt, the square of the distance between any two points on the plane must still be given by the Pythagorean theorem. In other words, the distance between points is a geometric invariant of the plane and therefore cannot be altered by any arbitrary change of the coordinate system. Indeed, we can go further and conclude that if the distance formula for any two points, in terms of the coordinates of those points on the plane, is the Euclidean Pythagorean theorem in every possible coordinate system, then the geometry on the plane must be Euclidean.

The spherical surface (the surface of the Earth) is an example of a surface on which the standard Pythagorean theorem holds in a Cartesian coordinate system only in a small area in the neighorhood of a point; it breaks down if we try to extend it to large distances. Instead of Cartesian coordinates we must use systems of circles (longitude and latitude) to construct a coordinate system. From this discussion, we thus conclude that the geometry on the Earth is non-Euclidean.

To study the laws of nature we must introduce a three-dimensional coordinate system since the position of an event in space can be uniquely specified only if its distances from three mutually perpendicular lines that intersect it at a point (the origin) are given. We call these distances the three coordinates $x$, $y$, $z$ of the event in space. Again, we may choose any kind of three-dimensional coordinate system we please, and again we impose on the laws of nature the principle of invariance; statements are laws only if they remain unchanged when we change our coordinate system.

The invariance of certain aspects of the laws of nature to a change of coordinate systems reveals important symmetries in nature that are related to the invariant features of these laws. To investigate these invariance–symmetry relationships, we first define the symmetry concept in science, then introduce the entities that are invariant to a transformation (change) of the coordinate system, and, finally, see how invariance and symmetry are related. We are all acquainted with the geometrical symmetry (or lack of it) of objects all around us and we generally associate such symmetry with beauty. The sphere has perfect symmetry in the sense that it looks the same no matter how we view it; its symmetry is called spherical symmetry and, as far as we know, space is spherically symmetric. We can now carry the symmetry concept over to mathematics. In algebra, the equation $y = x^2$ plotted in the $x$-$y$ plane is symmetric with respect to the $y$ axis: it is the same on the left side of the $y$ axis as on the right side. The ellipse also has symmetry with respect to the $x$ and $y$ axes; its shape below the $x$ axis is the mirror image of its shape above the $x$ axis and its shape to the right of the $y$ axis is the mirror image of its shape to the left of the $y$ axis. We can give many other examples of such symmetry but these are sufficient to define symmetry in the shapes of objects and in mathematics.

We now consider the symmetry of the laws of nature. If such laws are expressed mathematically, then these laws possess the symmetry of the mathematical formulas that express them. In addition to the mathematical symmetries that are imposed on the laws by the formulas that express them, the laws have certain dynamical features called conservation principles that imply other symmetries. To examine this idea we note that nature is a strict double-entry bookkeeper when certain dynamical entities in the motions of bodies are involved. Three of these entities are of special importance: the momentum of a body, its energy, and its rotational motion. If $m$ is the mass of a particle (the amount of matter it contains) and $v$ is its velocity, then its momentum is defined as the product $mv$. In this product, $v$ specifies not only the speed of the particle but also the direction of its motion. The particle carries its momentum $mv$ with it, and retains this momentum only so long as it is not interacting with other particles. In interacting with another particle, the initial particle may gain some momentum or lose some of its own, but any change occurring in the momentum of this initial particle is exactly balanced by an opposite change in the momentum of the second interacting particle so that the total momentum, taking into account both interacting particles, remains unaltered. In a collection of interacting particles, momentum may be transferred from particle to particle but the total amount of momentum can never change. This is called the principle of conservation of momentum. The product $(1/2)mv^2$ is the kinetic energy of the particle. This kinetic energy can also pass from particle to particle in an ensemble of interacting particles. But the total amount of energy does not change; this is the principle of conservation of energy. Finally, particles can be spinning or revolving around each other. This spinning or rotational motion is called angular momentum and it, too, is governed by a conservation principle in the sense that it can be transferred from particle to particle but the total amount is always the same.

Each of these conservation principles is related to or a consequence of an invariance with respect to a change of the coordinate system. If the laws of nature do not change when we transfer our coordinate system from one point of space to another point, the principle of the conservation of momentum is obeyed. If the laws do not change from moment to moment (the laws are invariant to the flow of time), the principle of the conservation of energy is obeyed, and finally, if the laws are the same no matter how we orient our coordinate system (the laws are invariant to a rotation of the coordinate system), the principle of conservation of rotation or angular momentum is obeyed.

From this discussion we see how intimately the mathematical formulation of the laws of nature (their mathematical symmetries) are related to basic dynamical principles. Thus, the mathematics becomes part of the laws; only through mathematics can we deduce all the consequences of the laws. This is a consequence of the following remarkable and powerful property of mathematics as a tool of the theoretical physicist. The mathematical manipulation of any law (expressed mathematically) does not alter the truth of that law and any mathematical deduction from the law so obtained is as valid as the law. This has enabled scientists to deduce truths about nature that go far beyond the truths that are immediately apparent in the laws themselves. Indeed, this is the principal activity of the theoreticians today—the mathematical manipulation of the basic laws to construct new truths. This is a game that anyone knowing the basic laws and adept at mathematics can play.

As we have already noted, we are forced to assume a particular geometry for our three-dimensional space. The laws of classical physics are based on Euclidean geometry (flat three-dimensional space) and the assumption that space and time are absolute and unrelated to each other. The physics of Einstein replaced classical physics in two ways. First, the absoluteness of space and time and their independence of each other had to be discarded and replaced by a four-dimensional space-time manifold. Second, the Euclidean geometry had to be replaced by a non-Euclidean four-dimensional geometry (a curved four-dimensional space-time manifold). This change required a special kind of mathematics from which it became clear that the laws of nature are part of geometry itself, and vice versa.

We come now to the roles of the differential and integral calculus in the remarkable development of science from the time of Newton to the present. The discovery of the basic laws of science are but the first phase in this great intellectual adventure. The second phase is using these laws to construct models of the structures in the universe from the nuclei of atoms to stars, galaxies, and the universe itself. The laws are expressed in differential form, which means that the differential calculus is used in expressing these laws. To deduce or build a structure (a model) of an atom or a star or a planetary orbit means putting the elements (differentials) together or integrating to go from the infinitesimal to the whole.

We illustrate this idea by describing briefly how Newton's classical laws, expressed in differential form (in terms of infinitesimals), led to the correct orbits of the planets (an integration operation). To obtain or deduce the orbit around the sun of a planet like the Earth we must know two laws, both of which were discovered by Newton: the law of motion and the law of force. The law of motion is a general statement in the form of an algebraic equation which states how the motion of a body (its velocity) is affected by a force acting on it. Newton, basing his conclusions on Galileo's experiments, formulated his three laws of motion which state, essentially, that a body's state of motion can be changed only by a force acting on the body. Unless a force acts on the body, it continues in the same state of motion (rest or constant speed in the same straight line). This statement is made precise by an algebraic equation, which equates the force $F$ acting on the body to the product of the mass of the body and the rate at which its velocity is changing, owing to the action of the force. Labeling the rate at which the velocity is changing with the letter $a$ (its acceleration) and its mass with the letter $m$, the algebraic equation, expressing Newton's law of motion (perhaps the most famous equation in the history of science), is $F = ma$. All of classical physics stems from this simple equation.

How does the differential calculus enter into this statement of the law of motion? To see why the calculus is required to write this law in a form which can be applied to the motion of a body, however complex it may be, we consider in detail the motion of a planet in its orbit around the sun. As the planet moves, the force acting on it changes from moment to moment. Since this is not shown by the equation $F = ma$, we must replace it by its counterpart for the instantaneous motion of the planet—that is, for its motion during an infinitesimal time interval $dt$. The differential calculus does this for us. To obtain what we want, we note that since the acceleration $a$ is the time rate of change of the velocity $v$ of the planet, we have $a = dv/dt$ or, in Newton's notation, $a = \dot{v}$. The law of motion for the planet therefore becomes $F = m\,dv/dt = m\dot{v}$, where the force $F$ is a function of the distance of the planet from the sun. As the equation stands, it can be applied to the motion of any object which is subject to any force $F$. To apply it to a planet we must specify the force that the sun exerts on the planet, which is the gravitational force.

Merely stating that $F$ is the gravitational force does not help us; we must express $F$ as a function of the distance $r$ of the planet from the sun and as a function of anything else it may depend on. Newton discovered the algebraic form of this function $F$. He demonstrated that $F$ depends on the product of the mass $M$ of the sun and the mass $m$ of the planet divided by the square of the distance $r$ between the sun and planet (Newton's law of gravity): $F = GMm/r^2$, where $G$ is a universal constant. With this expression for $F$, we can now write the equation of motion of the planet as $GMm/r^2 = m\,dv/dt$. But $v = dr/dt = \dot{r}$, and $dv/dt = d/dt(dr/dt) = d^2r/dt^2$. The equation of motion thus becomes $GMm/r^2 = m\,d^2r/dt^2$ or $GM/r^2 = d^2r/dt^2$ since the factor $m$ cancels out. This differential equation, as it is called, specifies the acceleration of the planet (the right-hand side of the equation—the time rate of change of the time rate of the position $r$ of the planet relative to the sun) in terms of $r$.

Determining the nature of the planet's orbit therefore necessitates that we find $r$ using our knowledge of its second derivative (acceleration) for each value of $r$, which is exactly what the equation of motion tells us. This is clearly the inverse of differentiation and is therefore an integration process; it is called "solving the differential equations of motion." It is remarkable that the solution of this simple equation, which can be found in a matter of minutes, gives us the three laws of planetary motion that Kepler deduced empirically from the observational data of Tycho Brahe after some thirty years of extremely laborious calculations. The solution gives us elliptical orbits for the planets and shows us that an orbit is completely determined by two adjustable parameters (constants): the semimajor axis $a$ of the orbit (the size of the orbit) and its eccentricity $e$ (the shape of the orbit). Moreover, the solution also shows that these two geometric constants are related to the two basic dynamical constants of the planet's motion: its energy and its rotational motion or angular momentum; $a$ determines the energy and $e$ determines the angular momentum for the given size $a$ of the orbit. This is one of the most beautiful examples in the history of science of the intimate relationship between science and mathematics.

# Epilogue

The preceding chapters span many centuries in the history of the most remarkable creation of the human mind. Mathematics stands above all other mental edifices in that it is a boundless web of interwoven threads of pure thought. The most notable feature of this web is the interrelationships of the threads that constitute it; if we start from any single thread we can reconstruct the entire web by abstract reasoning without reference to or the use of anything but the few rules that we may use to interlace these threads of thought.

Although arithmetic, the basis of all mathematics, started not as a purely intellectual discipline but as a mental device for keeping records of transactions involving the exchange of quantities of things and the recording of various kinds of measurements, its subsequent development had little to do with any practical application. Indeed, if arithmeticians had concerned themselves only with using arithmetic as a computational tool, one of the most beautiful and subtle branches of mathematics, "the theory of numbers," would never have been developed. This is also true of complex numbers, which are the foundation of a branch of advanced mathematics, "the theory of functions of complex variables," that has attracted the greatest mathematicians since its emergence in the mid-18th century. We note in passing that "the theory of numbers" boasts such problems as Fermat's last theorem, which have defied mathematicians for centuries. Trying to prove Fermat's last theorem is something of a "rite of passage" which every young aspiring mathematician accepts.

Even though arithmetic is the basic and most elementary branch of mathematics, it does not appear to demand anything more for its pursuit than computational skill. This is a very restrictive view of arithmetic. The thorough analysis of even simple problems in arithmetic may require the application of advanced mathematics. A striking example is that of the distribution of prime numbers. The solution of this problem lies in finding a general formula which tells us the number of primes that lie in any given numerical interval. In the early part of the 20th century, the great mathematician Edmond Landau, who taught at the University of Göttingen, wrote two large volumes analyzing this problem without solving it, using the most advanced mathematics known at that time. Even in the elementary aspects of mathematics we are thus dealing with complex topics which make great demands on our mathematical skills. Since most people find mathematics somewhat forbidding, if not frightening, they find it difficult to understand how it can be regarded as beautiful. We believe that this beauty will be revealed to those who read this book carefully. It is not the visual beauty of a painting or the audio beauty of a musical performance. Nor is it the literary beauty of a great poem; it is entirely intellectual and therefore, while more difficult to perceive, more satisfying when perceived.

The beauty to which we respond in mathematics is related to symmetry, which is the basis of all beauty whether it is visual, auditory, or literary. But symmetry itself is essentially a mathematical concept, which we discussed briefly in this book. The most beautiful symmetries that occur in nature are those associated with crystals such as the various precious stones (diamond, ruby, etc.). Each of these structures can be described by a definite mathematical formula so that the mathematician perceives the beauty of the precious stone in the mathematical formulas that represent it, just as the musician hears the sounds of a composition when he peruses its sheet music. The most beautiful manifestation of symmetry in nature that is immediately visible to the naked eye is that of the snowflake and one can express it by a very simple mathematical formula.

We come now to the role that mathematics plays in filling the human need for intellectual problems. Many of us satisfy this need by playing such games as chess and bridge or by doing various kinds of puzzles such as crossword puzzles but nothing can present us with the wealth of problems posed by mathematics. Humanity's penchant for problem-solving probably led to the emergence of mathematics, which then developed as a purely mental exercise, until its usefulness in surveying, navigation, construction, and commercial transactions became evident and it was accepted as a practical discipline to be taught in schools. But to professional mathematicians this was a minor and relatively unimportant phase in the growth of mathematics to which they devoted very little time. Mathematics thus branched out along abstract lines that appeared to have little application or relevance to practical problems. This trend continued until Kepler's discovery that the orbits of the planets around the sun can be described by the same mathematical equations that describe conic sections. This remarkable relationship between a purely abstract mathematical concept, the conic section, and a very practical (physical) entity, the orbit of a planet, seemed magical. Kepler was puzzled by it but could find no reason for this startling convergence of fact and theory. It remained for Newton to discover this reason by showing that the mathematical formula of the orbit of a planet is a direct consequence of the mathematical formula for the sun's gravitational force on a planet. This purely intellectual deduction of the observed orbit of a planet from the mathematical formula for the force of gravity (the Newtonian law of gravity) was a watershed in the development of both mathematics and physics. It demonstrated the deductive power one could achieve by combining a mathematical law of nature with correct mathematical reasoning.

Until Newton's discovery of the mathematical formula for the law of gravity, mathematics and physics had evolved separately, with their paths crossing only occasionally and just fortuitously. All that changed when Newton deduced Kepler's three laws of planetary motion from his law of gravity by pure mathematical reasoning. Science and mathematics then became inextricably interlinked, forming an interdependent pair that fed upon each other (a kind of symbiosis) in a most remarkable way. Indeed, mathematics divided physicists into two groups, the experimentalists and the theoreticians, who were also known as "mathematical physicists." The mathematical physicists (astronomers and chemists are included) not only used mathematics to solve the immediate unsolved problems in their discipline, but greatly influenced the growth and the development of mathematics. Mathematicians quickly accepted their subsidiary role and began reconstructing the mathematical forms of the laws of physics—particularly Newton's laws of motion and gravity—ultimately casting them into forms which greatly enhanced their usefulness and revealed physical features of gravity and dynamics that were not easily apparent in the simple forms of the laws as Newton presented them. Much of this work was done by the great mathematicians of the 18th century such as Lagrange, Poisson, d'Alembert, Fermat, and Maupertuis. This work was continued in the 19th century by such intellectual giants as Gauss, Hamilton, and Jacobi. By that time the interrelationship between the two disciplines had become so close that it was, in many instances, impossible to distinguish between mathematics and mathematical physics. Indeed, the mathematicians mentioned above thought of themselves as physicists as well as mathematicians.

Although physics had a great impact on mathematics, the reverse was and still is true to a greater extent. The reason for this is that we may manipulate the mathematical formulas that express the basic laws of physics, such as the laws of motion of bodies and the law of gravity, any way we please, consistent with the rules of mathematics. These manipulations make it possible for us to obtain new mathematical relationships among the bodies that reveal properties of their motion (their orbits, for example) that cannot be directly observed in the original equations. Such manipulations, called "solving the equations of motion," have been developed to such a point by mathematicians of the last two centuries that they constitute an entire branch of mathematics of their own which every physicist must master.

But mathematics has played another very important role in physics which could not have been foreseen; it has been the creator of new physical entities that began as purely intellectual concepts and then acquired deep physical meanings and lives of their own which greatly enriched and illuminated basic physical principles. A very important example of this magical creativity of mathematics is the discovery, or, better, the introduction into dynamics, of the concept of energy. The Newtonian equations of motion of a planet around the sun show that, as the planet moves from point to point in its orbit, a certain mathematical sum of a positive quantity and a negative quantity, both associated with the motion of the planet, remains unaltered. Both terms change as the planet moves, but the sum remains constant. If the magnitude (numerical value) of the positive term increases, the magnitude (numerical value) of the negative term increases by exactly the same amount so that the sum is unaltered. Since the positive term is the product of the mass of the planet and its speed squared, physicists called it the planet's kinetic energy and they defined the negative term, which depends on the mass of the planet, the mass of the sun, and the planet's distance from the sun, as the potential energy of the planet in the sun's gravitational field. The sum of these two different kinds of energies is called the total dynamical energy of the planet. The mathematical deduction that this total energy is the same no matter where the planet is in its orbit led to one of the most profound and far-reaching principles in all of science—the principle of the conservation of energy. The most remarkable aspect of this purely mathematical discovery of the abstract concept of energy is that it quickly acquired a physical, concrete reality that has to be taken into account in analyzing the dynamics of particles of matter in the matter itself.

The mathematical analysis of the dynamics of planetary motion led to another important deduction which also became a general principle which applies to all systems of interacting bodies whether they are the sun and planets or the electrons and nucleus in an atom. The mathematics shows that if the total dynamical energy of a body moving around the sun is positive (the numerical value of the kinetic energy exceeds the numerical value of the potential energy), the body (a planet, for example) cannot remain bound to the sun. The total energy of a system of bodies that are bound to each other must be negative. This rule applies to the stars in a galaxy, to the atoms and molecules in stars and planets, to the atoms in molecules, to the electrons and nuclei in atoms, and to the quarks in nuclei. The more negative this energy, the more highly bound is the system of particles.

That such a profound physical fact can be deduced from an essentially simple application of mathematics to the dynamics of bodies shows the tremendous intellectual power mathematics confers on the theoretician. But mathematics acquires a life of its own and ultimately controls the direction of the research pursued by the researcher; here the researcher (e.g., the theoretical physicist) must be aware of the danger of becoming a formalist if he surrenders too easily to the subtle persuasion of the clever and often beautiful mathematics that may have no physical interpretation at all. The discovery of "action" is an example of how mathematics can generate an entity that seemed to have no physical meaning initially but which is now accepted as an important physical property of the dynamics of particles. In the 18th century the mathematician Maupertuis was puzzled by how a planet "knew" the kind of path to follow in its motion around the sun. He rejected the concept of action at a distance (that gravity can act across the great distances between the sun and the planets) and proposed that the planet is guided around the sun by some property of the planet; he called this property, which is the product of the planet's mass, its velocity, and the distance it moves in a very short time, the "action" of the planet. The mathematics of the dynamics of the planet then shows that of all possible paths that connect two points of the planet's orbit, the planet chooses the one along which its action changes by the smallest amount.

Maupertuis was so enchanted with his remarkable discovery that he formulated it as a general physical principle which he called "the principle of least action" and assigned a theological interpretation to it. This purely mathematical concept was universally accepted by theoretical physicists and soon began to play a major role in the dynamics of interacting particles. The principle was later enlarged in the 1820s by the great Irish mathematician and mathematical physicist William Hamilton, and today it dominates theoretical physics even though "action" is still, physically speaking, a very mysterious entity which cannot be apprehended physically, even though it itself is a product of three easily understood physical entities. Here we see how mathematics becomes a physical reality, thus in a sense displacing the physics. Indeed it was easily shown mathematically that all dynamics can be deduced from the principle of least action.

With the introduction of action, physics became ever more dependent on mathematics for full understanding and so the theoretical physicists began to dominate the discipline because they were better mathematicians than the experimental physicists. Owing to the impact of mathematics on physics, the gap between the theoretician and the experimentalist today is greater than ever before.

The discoveries of the law of the conservation of energy and the principle of least action, direct consequences of applying mathematics to physics, were but the first in a series of such discoveries that enriched physics enormously and enabled physicists to penetrate deeply into matter to investigate the deep interiors of stars and to study the most distant regions of space. Mathematical physics also gave birth to new concepts so strange to the traditional physicist that the physics before 1900 is called "classical physics" and the physics after 1900 is called "modern physics." The genesis of modern physics is present in the papers of the mathematical physicist Hamilton (1820), the mathematician Gauss (1830), and the superb British theoretical physicist James Clerk Maxwell.

Hamilton showed that the mathematical equations that describe the motion of a particle in a varying gravitational field are similar to the equations that describe a ray of light passing through a medium of varying density. This discovery led to the conclusion that particles of matter and rays of light have certain physical properties in common. Gauss began the study and development of multidimensional non-Euclidean geometry which eventually led Einstein to his geometrical theory of gravity. Soon after Gauss, Maxwell did his astonishing theoretical work which was one of the greatest intellectual syntheses of physical concepts of all time.

When Maxwell began his theoretical work, the interrelationship of electricity and magnetism had been established experimentally by Faraday but the equations that had been derived to describe this relationship were not quite as symmetrical vis-à-vis the electric and magnetic fields as Maxwell thought they should be. To obtain equations with the fullest symmetry Maxwell therefore added another term which was a mathematical stroke of genius; it had no basis in any experimental evidence because it described or proposed a flow of an electric current in empty space, as though space itself were a conductor. This appeared to be an absurdity but the demands of symmetry could not be denied; the mathematics, in the hands of Maxwell, was there to supply it.

This single term revolutionized our understanding of electricity, magnetism, and light and immediately led to what we now call Maxwell's electromagnetic theory of light. From

Maxwell's symmetrical equations of the electromagnetic field one can derive, by fairly simple mathematical manipulations, a single equation that shows that an oscillating electric field traveling through space is always accompanied by an oscillating magnetic field and these two oscillations form a wave that has all the properties of light and travels at exactly the speed of light. Thus, mathematics revealed something that no one had ever expected and produced a technology of tremendous proportions.

Maxwell's electromagnetic equations mark the peak of classical physics or what we may call the Newtonian era since none of the mathematical developments up to that point questioned or challenged the Newtonian concepts of absolute space and absolute time, but physics outgrew these concepts, again with the help of mathematics, in such a revolutionary manner and with such a revolutionary impact that some physicists and many philosophers still do not accept these developments which constitute modern physics. Two great theories—the quantum theory and the theory of relativity—neither of which could have been constructed in their most advanced forms without fairly sophisticated mathematics—form the foundation of contemporary physics. But both of these remarkably beautiful and subtle theories can be understood fairly well without mastery or even knowledge of the mathematics that reveal them.

The quantum theory was born from the vain attempt to obtain a mathematical formula, based on classical physics, that correctly describes, in all its details, the radiation emitted per second from one square centimeter of a furnace at a given temperature. We all know that a hot furnace, like the sun, emits radiation which is a mixture of all colors, from red to violet, but the colors in this mixture are not equally intense; the temperature of the furnace determines the intensities of these various colors, which, of course, can be measured quite easily. Near the end of the 19th century, theoretical physicists tried to derive a mathematical formula from which one could calculate these intensities. One of the physicists who was most active in the search for this formula was the great German theoretician Max Planck.

Since radiation, whether from the sun or from a furnace, is an electromagnetic phenomenon, Planck looked for the formula in some mathematical aspect of Maxwell's electromagnetic theory of radiation. But he also had to take into account the laws of thermodynamics (heat) since the formula he was seeking also had to contain the temperature. But however hard he tried, he could not derive the formula unless he introduced a very revolutionary hypothesis that rejected a basic principle of Newtonian and Maxwellian physics: the principle of continuity which states that if an event A leads to an event B (A is the cause of B) then these two events are connected by an unbroken chain of events which can be followed in as great detail as we wish.

Though this principle seems reasonable, Planck had to discard it before he could derive the correct mathematical formula that describes the furnace radiation. He had to assume, in fact, that this radiation is not emitted as a continuous wave but rather in the form of small bunches of radiation, like bullets from a machine gun. He called these bundles of energy "quanta" and so the "quantum theory" was born. This theory was a direct consequence of the need to express the rate at which the furnace emits radiation as a mathematical formula; here mathematics pointed to the correct result, forcing Planck to replace the concept of continuity, which he cherished, by one that he found repugnant.

The discovery of the second of the two great theories—the theory of relativity—that guides us in our search for an understanding of the universe also owes a great deal to mathematics. Indeed, the development of the special and general theory of relativity was primarily a mathematical synthesis of space, time, and geometry which forever altered our understanding of the laws of nature and established a universal criterion for determining the truth or falsity of what may or may not be a law.

The special theory itself stemmed from the observation that the speed of light in a vacuum is the same for all observers, regardless of how fast (or slowly) these observers are moving with respect to each other. Incorporating this crucial fact in his mathematical analysis of how these different observers interpret physical events, Einstein showed three-dimensional space and one-dimensional time are interwoven by the mathematics into a single four-dimensional space-time manifold. All the wonderful consequences of the special theory of relativity, such as the equivalence of mass and energy, can be deduced mathematically from this constancy of the speed of light alone. Even more wonderful was Einstein's magnificent use of non-Euclidean geometry to replace the concept of gravity as a force by the concept of gravity as a distortion (curvature) of space-time. Here the power of mathematics to expand our knowledge about the universe is obvious.

The next important phase in this remarkable story of the interdependence of physics and mathematics which came in the 1920s stemmed from the work of Niels Bohr. He showed mathematically that the application of the quantum theory to the motions of electrons in atoms explain why and how these atoms emit the radiation they do when they are stimulated (excited). Despite the brilliance of the Bohr theory of the atom and the simplicity of the algebra that Bohr used to construct it, the theory is basically flawed because it is a mixture (hybrid) of Newtonian continuous mechanics and the discontinuities of the quantum theory and so physicists began looking for a mathematical model of the atom that does not suffer from this duplicity of two antagonistic principles. For some ten years following Bohr's work, physicists did not know how to produce such a model and then a number of exciting mathematical deductions were made by four young theoretical physicists independently which led to an entirely new mechanics—the quantum mechanics—that provided the necessary clues. The theoretical physicist Louis de Broglie took the first step in this development. He showed that particles, like electrons, have wave properties as well as particle properties. This discovery led to the wave–particle dualism concept that now dominates physics: electromagnetic waves have particle features (photons) and all material particles have wave features (de Broglie waves).

The next step in the development of this remarkable chapter of modern physics was taken by the theoretician Werner Heisenberg who discovered what is now called the "uncertainty principle," which limits the accuracy with which we can know simultaneously both the position of a particle and how it is moving. One can show mathematically that this principle is equivalent to de Broglie's discovery of particle waves.

The third and most important contribution to this construction of the new quantum mechanics was made during this same period by the Austrian theoretical physicist Erwin Schrödinger; he discovered the wave equation that describes the wave motion of a particle such as an electron in an atom. This wave equation enables one to construct the de Broglie wave for any particle from which all properties of the electron can be deduced. In this analysis of the particle's motion, complex numbers must be used; the de Broglie wave that the solution of the Schrödinger equation gives for the electron is itself a complex quantity consisting of a real and imaginary part. Complex numbers thus play a very important role in the description of the motion of particles like electrons.

We come now to one of the most amazing discoveries in the history of science which stemmed from the mathematical combination of quantum mechanics and special relativity. The British theoretician Paul A. M. Dirac in the late 1920s was not entirely happy with the Schrödinger wave equation, in spite of all its successes, because the equation, as it stands, does not conform to the special theory of relativity which is a necessary constraint imposed on all theories. If a theory does conform to this constraint, it is described as being "relativistically invariant" (the same for all observers regardless of how fast they are moving with respect to each other). One can easily analyze any theory mathematically to check whether or not it is relativistically invariant. If it is not relativistically invariant, it is not an easy mathematical task to make it relativistically invariant although it is not impossible. Dirac did this for the wave equation of the electron and thus, with great brilliance, constructed the "relativistic wave equation of the electron."

This equation (really a combination of four equations) solved many subtle atomic problems which the Schrödinger equation cannot solve, but more important than this expansion of our knowledge of the electron and the atom which the Dirac equation gives, is its prediction of the existence of antimatter, a purely mathematical deduction from the Dirac equation which Dirac himself obtained. This prediction was verified a few years later when the experimental physicist Carl Anderson discovered positrons (antielectrons) among the cosmic rays that entered his cosmic ray detector. This is a magnificent example of the predictive power of mathematics; the prediction of antimatter could not have been achieved without mathematics.

In this epilog we have touched only on some of the high points of the great contributions of mathematics to science; the reader can enlarge this list of such contributions from his own knowledge. As science and technology progress, the increasing complexities this progress produces require for their study increasingly more complex mathematics. Indeed, science and technology could not have reached their present state without mathematics. Not so long ago, mathematicians used to consider the very abstract branches of mathematics such as number theory as totally pure and having no use in physics or technology. But this belief is wrong. Physicists use almost all branches of mathematics. Physics and mathematics are thus coupled in a never-ending dance which may produce wonders we cannot even dream of now. It is exciting and, in a very abstract sense, beautiful that two intellectual disciplines, one probing the universe and the other constructing pure thought, without reference to the universe, can produce such magical results.

# Index